The Design of Electrical Services for Buildings

The Design of Electrical Services for Buildings

F. PORGES

B.Sc.(Eng), F.I.Mech.E., M.I.E.E.

John Porges, Consulting Engineers

SECOND EDITION

LONDON   NEW YORK
E. & F. N. SPON

First published 1974
by Chapman and Hall Ltd
11 New Fetter Lane, London EC4P 4EE
Second edition 1982
Published in the USA
by Chapman and Hall
in association with Methuen, Inc.
733 Third Avenue, New York NY 10017
© 1974, 1982, F. Porges

Printed in Great Britain by
J. W. Arrowsmith Ltd, Bristol

ISBN 0 419 12360 1 (cased)
ISBN 0 419 12370 9 (paperback)

British Library Cataloguing in Publication Data

Porges, F.
    The design of electrical services for buildings. —
    2nd ed.
    1. Buildings — Electric equipment
    I. Title
    621.319'24   TK4001   60862

    ISBN 0-419-12360-1
    ISBN 0-419-12370-9 Pbk

Contents

Preface to Second Edition

This book sets out to provide a basic grounding in the Design of Electrical Services for Buildings. It is intended for students of building services engineering in Universities and Polytechnics and will also be useful to graduates in mechanical and electrical engineering who are about to specialize in building services after obtaining a more broadly based first degree. The emphasis throughout is on the needs of a design engineer rather than those of an installation electrician or of an architect.

In spite of the increasing number of specialized first degree courses now in existence, engineering is one discipline and I believe it is the colleges providing a general course in mechanical or electrical engineering which will continue to produce our best engineers. These young men and women learn to apply their knowledge during their practical training and during their first jobs. They can learn more quickly and more easily if they have some guidance at this stage of their careers but I have always found a lack of books which would bridge the gap between a theoretical text and the unwritten experience of one's seniors. It was in the hope that I could meet this need that I originally wrote this book.

Opinions will always differ about the order in which the topics within the subject should be taken. On the assumption that I am writing for someone with no previous knowledge, I felt that I should describe the equipment used before explaining how the electricity is distributed to it. I personally would find it confusing to try to explain distribution without first saying to what the supply has to be distributed. Those who find a different order clearer may prefer to read the chapters out of sequence.

In this edition examples of calculations have been added and a new chapter has been included which describes a complete design of an industrial system. This chapter concludes with a typical specification which will show students how theoretical knowledge is applied in practice. I have also added a bibliography and each chapter now has a list of British Standards and those sections of the I.E.E. Regulations which are relevant to it.

The chapter on protection has been considerably rewritten to make the presentation clearer and, also, to bring it into conformity with the 15th Edition of the I.E.E. Regulations.

I have included a chapter to explain the form and function of the I.E.E. Regulations but have not attempted any commentary on them. The intention of this book is to provide something more than a gloss on the regulations. A book which hopes to cover the complete design of an electric installation must include many things not dealt with by regulations and should be free to follow its own methods and sequence. Once this was done there was nothing to be gained by covering the same ground a second time in the form of a commentary or explanation of the regulations.

The subject matter of this book is the design of electrical services in buildings and I have kept strictly to this. There are in practice many cases where the electrical designer relies on information and assistance from specialists in related but separate fields. This applies in particular to controls for heating and air conditioning, which are designed by specialists in that field and not by the consultant or contractor employed for the general electrical system. A description of them would therefore be out of place here. Many other services within a building include electrical equipment but the principles of motors, thermostats and controls are major studies of their own. Electric heating undoubtedly uses electricity but its design requires a knowledge of heating and ventilating. All these are topics which embrace more than the purely electrical work within a building and if they are to be dealt with properly they must have books of their own. Whilst appreciating that they may well form part of a complete engineering course I do not think they can all be covered in one book, and rather than treat them superficially and incompletely I have left them out altogether.

I must again thank the firms and organizations which have lent or given photographs for illustrations and to Mr P. Read of E. & F. N. Spon for his help and guidance with the procedure of revising the book.

London   F. P.

1 Accessories

Introduction

From the user's point of view the electricity service in a building consists of light switches, power points, clock connectors, cooker control units and similar outlets. Such fittings are collectively known as accessories; this name came about because they are accessory to the wiring, which is the main substance of the installation from the designer's and installer's point of view. To them, the way the outlets are served is the major interest, but it is quite secondary to the user who is concerned only with the appearance and function of the outlet. In the complete electrical installation of a building the wiring and accessories are interdependent and neither can be fully understood without the other, but a start has to be made somewhere, and in this book it is proposed to consider accessories first.

Switches

A switch is used to make or interrupt a circuit. Normally when one talks of switches one has in mind light switches which turn lights on and off. A complete switch consists of three parts. There is the mechanism itself, a box containing it, and a front plate over it.

The box is fixed to the wall, and the wires going to the switch are drawn into the box. After this the wires are connected to the mechanism. To carry out this operation the electrician must pull the wires away from the wall sufficiently to give himself room to work on the back of the mechanism. He then pushes the mechanism back into the box and the length of wire which he had to pull out from the wall becomes slack inside the box. It is therefore important that the box is large enough to accommodate a certain amount of slack wire at the back of the mechanism.

Standard boxes for recessing within a wall are 35 mm deep. There are also shallow boxes available which are 25 mm deep. Sometimes the wiring is done,

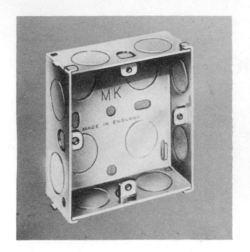
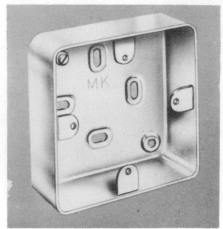

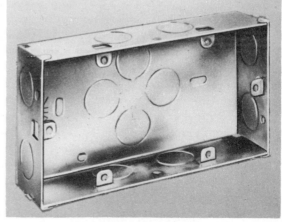

Fig. 1 Boxes (*Courtesy of* M.K. Electric Ltd.)

not in the depth of the structural wall, but within the thickness of the plaster. For use with such wiring boxes are made 16 mm deep. It is often necessary to install wiring and accessories exposed on the surface of wall. For such applications surface boxes are made which are both more robust and neater in appearance than boxes which are to be recessed in walls and made flush with the surface. Typical boxes of both types are shown in Fig. 1.

The older type of switch mechanism was dolly operated. It is illustrated in Fig. 2. The moving contact was on a spring lever which was moved by a cam. The cam was attached to the switch dolly. The shape of the moving contact and of the cam ensured that the contact snapped into either its on or its off position, according to the position of the dolly, but would not stay in an intermediate position.

Dolly operated switches have now been entirely superseded by rocker operated switches. This type is also illustrated in Fig. 2. It has a rocker which is pivoted at its centre and which carries a spring loaded ball. The ball presses on the moving contact and the combination acts like a toggle; the spring always forces the moving contact into one of its two extreme positions. The switch opens when the bottom of the rocker is pressed and shuts when the top is pressed. The advantages of the rocker switch are that it is easier to operate and that it is almost impossible to hold it half open, even deliberately. The disadvantages are that it is not so easy to see at a glance whether it is on or off and that it is more easily switched from one position to the other by an accidental knock.

There is a maximum current which the contacts of any particular switch can make or break, and a switch must not be put in a circuit which carries a current greater than that which the switch can break. Most manufacturers make switches in two standard capacities, the lower being rated at 5 amps and the higher at 15 or 20 amps according to the manufacturer.

Fluorescent lights have an inductive load, and the voltage surge which occurs when an inductive load is broken must be taken into account in selecting a switch for fluorescent lighting. It was for this reason that some of the older switches had to be de-rated when they were used for fluorescent lights, but switches in current production are suitable for inductive loads up to their nominal rating.

5 A is not as large a rating as one might think at first sight. If ten tungsten lamps of 100 W each are controlled from one point the total current to be switched is 4.2 A. Similarly seven twin tube 5 ft 65 W fluorescent lights take a total current of 5.2 A. Combinations such as these can easily occur in public buildings and it is often advisable to use 15 A switches.

When the switch is wired and inserted in its box it needs a front plate over it. This is often a loose component with a hole which fits over the dolly or rocker and which is screwed to lugs on the box. Standard boxes always have lugs for that purpose. A switch with a separate front plate is called a grid switch. Alternatively the switch may be a plate switch in which case the front plate is made as

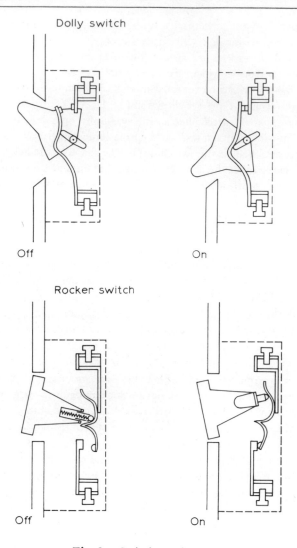

Fig. 2   Switch mechanisms

part of the switch and not as a separate piece. Both plate and grid switches are illustrated in Fig. 3.

Grid switches are so called because with this type several mechanisms can be assembled on a special steel grid. This makes it possible for banks of any number of switches to be made up from individual mechanisms. Standard grids and front plates are available for almost any combination which may be required, and special boxes to take these assemblies are also available.

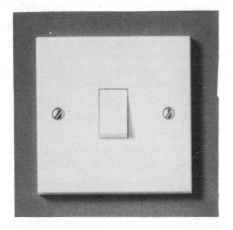

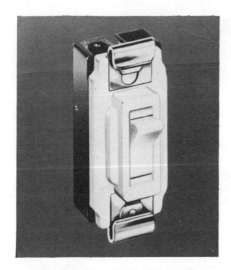

Fig. 3   Switches (*Courtesy of* M.K. Electric Ltd.)

The standard switch boxes described so far are intended either to be fixed on a wall or to be recessed in it. Narrow boxes and switches are also made which can be recessed within the width of the architrave of a door. These are known as architrave switches. The grid switch shown in Fig. 3 is of the architrave pattern.

Another type of switch is made which has no protruding lever or rocker, but is operated by a key which has to be inserted into the switch. This type of switch is very useful for schools and the public areas of blocks of flats. The caretaker has a key with which he can operate the lights but unauthorized persons cannot turn lights on or off.

Safety regulations often make it impossible to use ordinary switches in bathrooms. For such situations ceiling switches are made which are operated by an insulating cord hanging from the switch. The switch itself is on the ceiling and the cord hangs down to normal switch height.

In order that power equipment can be fully isolated it is often desirable to use a double pole switch. This expression means a switch which opens both the live and neutral circuits. The mechanism is similar to that of an ordinary, or single pole switch, but there are two contacts working side by side. The only difference visible on the outside is that the switch is larger and heavier. A double pole switch is shown in Fig. 4. This particular one is surface mounting and has a neon indicator, but double pole switches are also made without neons and for putting in recessed boxes.

Fig. 4   Double pole switch (*Courtesy of* M.K. Electric Ltd.)

There are certain very common applications of switches such as water heaters and fans. Some manufacturers, therefore, make double pole switches with the words 'Heater', 'Fan', 'Bath' or whatever other use is envisaged engraved on the front plate.

The usual ratings of double pole switches are 15, 20, 30, 45 and 60 A.

Socket outlets

A socket outlet is the correct name of what is popularly known as a power point. It is a female socket connected to the power wiring in the building and will accept the male plug attached at the end of the flexible wire of an appliance such as a Hoover, electric fire or wireless.

The general arrangement of socket outlets is similar to that of switches. There is a box to house the outlet, the outlet itself and finally a front plate. In the case of socket outlets the front plate is usually integral with the outlet. In Great Britain the majority of socket outlets intended for domestic or commercial use are designed for 13 A plugs. These plugs have three rectangular pins and the sockets have three corresponding rectangular slots to take the pins. Each plug also has a fuse inside it, so that each appliance has its own fuse at the feeding end of its trailing cable. This protects the appliance, and the fusing arrangements of the building wiring need protect only the permanent wiring of the building.

However, there are many older installations still in existence and plugs and sockets for use with them are still being manufactured. The older fittings all have round pins and sockets. They are rated at 2 A, 5 A and 15 A, the spacing of the pins and sockets being different for the different ratings. This makes sure that a plug of one rating cannot be inserted, even wilfully, into a socket of a different rating. Plugs and sockets rated at 2 and 5 A are available in both two- and three-pin versions, but those of 15 A rating are made only with three pins.

Two of the three pins are for the live and neutral wires, and the third one is for a separate earth wire. It should be noted that although a separate earth wire was not always provided on many older installations it is essential with all present day methods of wiring buildings.

Typical socket outlets are illustrated in Fig. 5. It will be seen that they are available with and without switches. Unswitched sockets have the contacts permanently connected to the wiring and are, therefore, permanently live. The appliance to be connected is turned on as soon as the plug is pushed into the socket, and is turned off when the plug is pulled out. If, however, a switch is incorporated in the socket outlet the switch must be turned on before the line contact becomes connected to the supply. The switch mechanisms built into socket outlets for this purpose are of the same types as those used for lighting switches. It is possible to leave a plug half in and half out of a socket so that parts of the bare pins are left exposed. If the socket is permanently live the exposed part of one of the pins is live and in this half way position it could be touched by a small finger or a piece of metal. Also if an appliance connected to the plug is faulty and takes an excessive current arcing can occur as the plug is pushed in and out.

These hazards are avoided if the socket is not switched on until after the plug has been pushed in. Of course there is nothing to stop a householder switching the socket on first and pushing the plug in afterwards, and in fact many people do this. The switched socket outlets in a house are then left permanently switched on, so that the advantage of a switch is lost. However, people will not learn to use equipment properly if they are not provided with it, and it may perhaps be regretted that unswitched sockets are made at all.

A further refinement to a socket outlet is the addition of a neon indicator light which shows when the socket is switched on. This can be reassuring to mechanically minded people who find electricity difficult and feel happier if something visible happens when a switch is turned on. It is also convenient for

seeing at a glance whether it is the power supply that has failed or the appliance connected to the plug which has developed a fault.

Like switches, socket outlets can be recessed into a wall with the front flush with the face of the wall or they can be mounted completely on the surface. The socket outlets illustrated in Fig. 5 are of both types.

Fused spur outlets

Fused spur outlets are used for connecting a single permanently fixed appliance to the wiring. They are used for example for connecting fixed as opposed to portable electric fires, water heaters and other equipment of this sort. Electrically, they perform the same function as a socket and plug combination, the difference being that the two parts cannot be separated as the plug and socket can. They must be used when a fixed appliance is to be served from a ring main circuit serving socket outlets as well as the fixed appliance. Fig. 6 shows some typical fused spur units.

Physically, they are similar to socket outlets and are connected to the wiring in the same way. They differ in that they have a fuse which is accessible for replacement from the front, and in that they have no sockets for a plug to be pushed into. The outlet connection is permanently wired, there being terminals for this purpose within the unit; the outlet cable is brought out of the unit either underneath or through the front. Like socket outlets, fused spur units can be switched or unswitched and can be with or without a neon indicator A switch on an outlet makes it easier to do repairs or make adjustments to the appliances served by the outlet. The disadvantages are that it costs a little more and that unauthorized persons may be able to turn the appliance on and off.

Shaver outlets

The use of shaver outlets is described in Chapter 9. The outlet itself consists of a two pin socket with a switch, the assembly being suitable for fitting into a standard box. Fig. 7 shows a shaver outlet which has the assembly on the back of the front plate and is suitable for fitting into a box recessed in the bathroom wall. Some shaver outlets are unswitched, in which case the sockets are permanently live as is the case with unswitched socket outlets. They are also available in switched versions with neon indicators.

Cooker control unit

Electric cookers take a much larger current than most other domestic appliances. They therefore require heavier switches than those used for lighting or in socket outlets. Moreover, it is usually convenient to have a socket outlet near the

Fig. 5   Socket outlets (*Courtesy of* M.K. Electric Ltd. and J.A. Crabtree & Co. Ltd.)

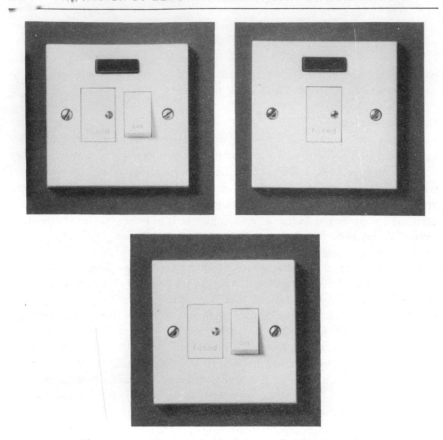

Fig. 6   Fused spur units (*Courtesy of* M.K. Electic Ltd.)

cooker in addition to the cooker switch itself. Cooker control units are, there-fore, made which have a 45 A (sometimes only a 30 A) switch with outgoing terminals for a permanent cable connection to the cooker and which also con-tain an ordinary 13 A switched socket outlet. The cooker switch is double pole, that is to say on opening, it disconnects both live and neutral lines, and the unit also has a substantial terminal for the earth wires.

A cooker control unit is shown in Fig. 8. Again units are available for both flush and surface fixing. The unit is mounted above and to the side of the cooker. The cable from the unit to the cooker is usually hidden in the wall and comes out at low level behind the cooker. A special flex outlet cover is made to fix on the surface of a box which is let in flush with the wall to make a neat outlet from the wall to the cooker. The flex outlet is normally supplied as a loose piece with the cooker control unit.

Fig. 7 Shaver outlet (*Courtesy of* M.K. Electric Ltd.)

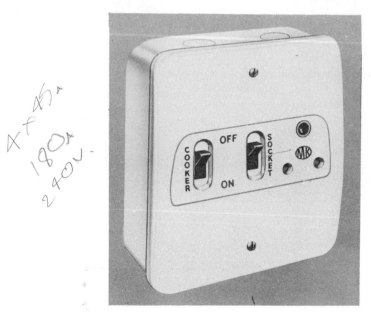

Fig. 8 Cooker control unit (*Courtesy of* M.K. Electric Ltd.)

Boxes

The use of boxes for housing switches and other accessories has already been described. The same boxes are used for conduit installations. When wiring is done by pulling cable through conduit, access must be provided into the conduit for pulling the cable in. Also where the paths of cables branch two or more conduits must be connected together. For both these reasons, a box of some sort is needed for use with conduit, and the type of box used is the same as that used for housing switches. As stated in the section on switches, boxes are available for recessing in walls, recessing within the narrow depth of plaster only or for fixing to the surface of walls. Where a large number of conduits is to be connected to the same box, the box is made longer in order to accommodate them side by side.

It can be seen in Fig. 1 that the boxes have a number of circles on them. These are called knock-outs and their circumference is indented to about half the thickness of the parent metal. It is therefore easy for the electrician on site to knock any one of them out in order to make a hole in the box. The hole so made is the right size to accept standard electrical conduit. It will be clear from the illustration that sufficient knock-outs are provided to make it possible to bring conduit into a box from any direction and in any position.

In addition to rectangular boxes of the sort illustrated, circular boxes are also made. These are useful for general conduit work and for terminating wiring at points which are to take light fittings.

When boxes are used for connecting lengths of conduit rather than for housing other accessories, they must have the open side covered with a blank plate. A typical plate is shown in Fig. 9. Circular plates are also made for circular boxes. It should perhaps not need saying that when a box is recessed in a wall the cover must be left flush with the surface of the wall so that it can be removed to give

Fig. 9   Cover plate for box (*Courtesy of* M.K. Electric Ltd.)

access to the cables inside the box. This is particularly important if the system is installed with the intention that it should be possible to rewire it later.

Many boxes have a wiring terminal which enables a wire to be connected to the metal of the box. This is used for connecting an earth wire. The purpose and use of earthing is discussed in Chapter 9. The importance of the earth terminal on the box arises when the accessory which is housed in the box has to be earthed through the box. This is particularly important when a plastic conduit system is used which necessitates the use of a separate earth wire. There must then be some means of connecting the earth wire to the accessory and this can become difficult if there is no suitable terminal in the box for making the connection. Not all boxes have such a terminal and some care must be exercised in ordering if it is known that an earth terminal will be needed.

TV outlets

Housing design today has to accept that every flat, maisonette or house will have a television which may require connection to an outdoor aerial. It is becoming increasingly common to provide a communal aerial system which serves all dwellings on an estate from a single aerial. The chief reason for doing this is that it avoids the ugliness of a large number of aerials, all of different patterns, put up close to each other by different people. It has the further advantage that one powerful aerial erected in a carefully chosen position can give better reception than the aerials which individual occupiers can put up.

If a communal aerial system is installed, it becomes necessary to run a television aerial cable from the aerial to an outlet in each dwelling. There has to be a suitable terminal in the dwelling, and this takes the form of a socket capable of accepting the coaxial plugs used on the end of aerial cable. An outlet of this kind is shown in Fig. 10. Since a television set also needs a power supply, it is usual to

Fig. 10   TV outlets (*Courtesy of* M.K. Electric Ltd. and Walsall Conduits Ltd.)

provide an ordinary socket outlet near the aerial outlet. One manufacturer makes a combined unit having an aerial socket and 13 A socket outlet within one housing, and this unit is also illustrated in Fig. 10.

For radios which require both an aerial and an earth connection, special two pin outlets are available. These can also be combined in a single unit containing the mains socket outlet as well as the two pin outlet.

(a)

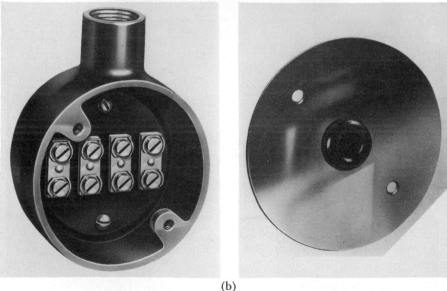

(b)

Fig. 11   Telephone outlets (Fig. 11a *by courtesy of* M.K. Electric Ltd.; Fig. 11b *by courtesy of* Walsall Conduits Ltd.)

Telephone outlets

To avoid the need for a lot of surface cable fixed after a building is occupied, it is quite common to put wiring for telephones in as part of the services built into the structure as the building is erected. This wiring must, of course, be brought to suitable terminals at the positions at which the telephones are to be connected later. The only essential requirement is an opening through which standard telephone cable can be brought out neatly. A plate with a suitable outlet which fits into a standard box, is shown in Fig. 11a.

A different telephone outlet is shown in Fig. 11b. This one consists of a four-pole connector with standard G.P.O. terminals, fitted in a standard circular box and a circular cover plate with a bushed opening to take a standard telephone cable.

Clock connector

Special outlets are made to which electric clocks can be easily connected. A typical one is shown in Fig. 12 and can perhaps be considered as a special purpose fused spur outlet.

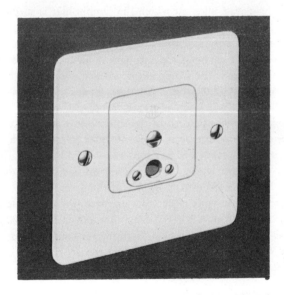

Fig. 12   Clock connector (*Courtesy of* M.K. Electric Ltd.)

It contains a 2 A fuse and terminals to which the wire from the clock can be connected. The fuse is needed because a clock outlet is usually connected to the nearest available lighting circuit. The fuse protecting the whole circuit will never be rated at less than 5 A, and may be as much as 15 A. The clock wiring is not suitable for such a large current and must, therefore, have its own protection at the point at which the supply to it branches from the main circuit. The necessary protection is provided by the fuse in the connector. The front of the connector has an opening through which the clock wire can be taken out to the clock. In most cases, the clock connector is made flush with the wall and the clock is subsequently fixed over it. However, surface connectors are available, and in this case the clock would be fixed next to the connector with a short length of wire run on the surface of the wall between the clock and connector.

Lampholders and ceiling roses

In public buildings the light fittings are fixed as part of the electrical installation. In housing, the choice of the lamp shade or fitting is usually left to the owner or tenant and is made after the dwelling is occupied. Plain lampholders are, therefore, provided which will accept ordinary 100 W and 150 W tungsten bulbs, and which usually have a ring to which a normal lampshade or similar fitting can be attached. The top of the lampholder screws down to grip the flexible wire cord on which it is suspended from the ceiling. Typical lampholders are shown in Fig. 13.

The flexible insulated wire on which the lampholder is suspended performs two functions. It carries the electric current to the lamp, and it supports the weight of the holders, lamp and shade. Its physical strength is, therefore, just as important as its current carrying capacity and it has to be selected with this in mind. At the ceiling itself, the wiring in (or on) the ceiling must be connected to the flexible wire. The connection is made by means of a ceiling rose, which is illustrated in Fig. 14. It consists of a circular plastic housing with a terminal block inside and a bushed opening on the underside where the flexible cable to the lampholder can come out of the rose. In installations which have the main wiring inside the ceiling, this wiring enters the rose through the back or top of the rose; when the main wiring runs exposed on the surface of the ceiling, it enters the rose through a cut out in the side of the rose.

Ceiling roses are made with either two or three line terminals in addition to an earth terminal. The reason for the third line terminal is explained in Chapter 5 and it will be seen there that when this third terminal is used, it remains live even when the light attached to the ceiling rose is off. It must, therefore, be shrouded so that it cannot be touched by accident if ever the flexible cord is being replaced.

In some situations, it is undesirable to have the lampholder hanging on the end of a flexible cable while there is no objection to having the lamp at ceiling height. In such cases, one makes use of a batten lampholder, which is illustrated in Fig. 15. It combines the terminal block of the ceiling rose with the lampholder in one fitting, and it can be screwed directly to a standard circular box on the ceiling. A batten lampholder could also be used to fix a light to a wall, but the lamp would project perpendicularly from the wall. The angled batten holder

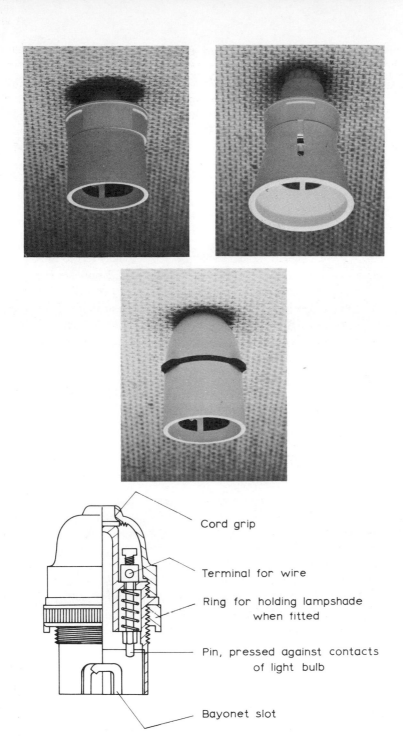

Cord grip

Terminal for wire

Ring for holding lampshade
when fitted

Pin, pressed against contacts
of light bulb

Bayonet slot

Fig. 13   Lampholders (*Courtesy of* Delta Electrical Accessories Ltd)

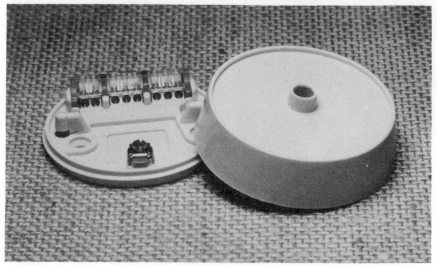

Fig. 14   Ceiling rose (*Courtesy of* Delta Electrical Accessories Ltd)

shown in Fig. 16 has the lampholder at an angle to the rose so that when the whole fitting is put on the wall the lamp is at a downward angle. Such angled battenholders can be obtained either with the lampholder at a fixed angle or with the angle adjustable.

Lampholders frequently have protective shields which are intended to prevent accidental contact with either metal parts or with the lampholder pins themselves. Lampholders with such shields are shown in Figs. 13 and 15. These

Fig. 15   Batten lampholder (*Courtesy of* Delta Electrical Accessories Ltd)

Fig. 16   Angled batten lampholder (*Courtesy of* Delta Electrical Accessories Ltd)

shields are often referred to as Home Office Skirts, and should always be used in bathrooms and kitchens where the atmosphere is likely to be damp.

Pattresses

It can happen that an outlet, such as a socket outlet or ceiling rose, has to be placed a small distance in front of the structure available to support it. This can happen, for example, when a wiring system is installed on the surface of walls and ceilings and there is a step in the surface which the wiring cannot follow so that it has to be supported off the surface. It is then necessary for some sort of distance piece to take up the gap between the fitting and the surface behind it. Standard components are available for this and are known as pattresses. A pattress for use with a circular socket outlet is shown in Fig. 17, which also gives a sketch showing the use of the pattress with surface conduit. The inclusion of the pattress makes it possible for the cables to enter the socket outlet from the back, whereas without it, there would be an untidy junction of the conduit with the bottom of the socket outlet.

Fig. 18 shows a pattress for use with circular surface switches and batten lampholders. Fig. 19 shows a different type of pattress. This is useful with some modern building methods in which the wiring is installed in a special skirting. The skirting is at floor level, but this is too low for socket outlets and the latter are, therefore, a little above the skirting so that at each outlet cables have to rise a small vertical distance. The pattress shown provides a neat and convenient way of doing this. It has also been known to happen that in the course of erection of

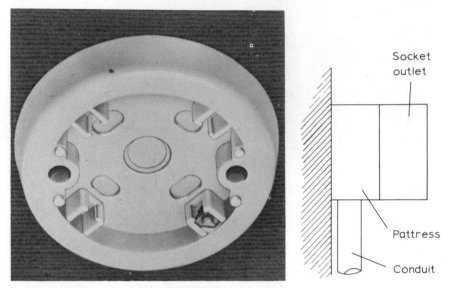

Fig. 17   Pattress (*Courtesy of* Delta Electrical Accessories Ltd)

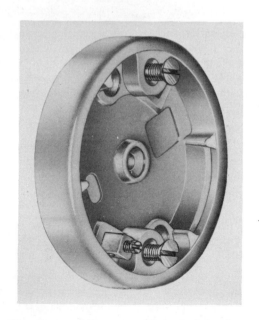

Fig. 18   Pattress (*Courtesy of* J.A. Crabtree & Co. Ltd.)

Fig. 19   Pattress (*Courtesy of* J.A. Crabtree & Co. Ltd.)

a new building an electrical outlet is wrongly placed. For example, a heating pipe to a radiator may run right in front of the box left to take a socket outlet. The type of pattress shown in Fig. 19 is a neat way of extending the wiring to an adjacent position, where the alternative might be to demolish large parts of a wall already built in order to give access to conduit buried in it, as the only means of extending that conduit.

Laboratories

Laboratories in schools, universities and industrial establishments often need special services which are not required in other areas. The commonest electrical service of this kind is a low voltage supply. This is normally provided by a transformer from the mains, which can either provide a fixed secondary voltage or be of the tap changing type to give a choice of voltages. Laboratory units are made specially for the latter purpose, and often incorporate a rectifier so that a low voltage d.c. supply is made available at the same time as a low voltage a.c. supply.

The low voltage is fed to outlets on the laboratory benches, and it is clearly

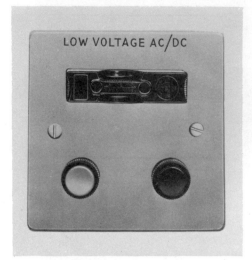

Fig. 20   L.V. outlets (*Courtesy of* The Wandsworth Electrical Mfg. Co. Ltd.)

important that these outlets should not be confused with the mains voltage outlets. Fig. 20 shows two low voltage units which have terminals to which laboratory equipment can be easily wired. One of them has a fuse; this has the merit that if there is a fault in the experiment connected to the outlet, the outlet fuse will blow and the supply to the other experiments will not be affected. Without this fuse, any fault would blow the fuse in the transformer secondary and thus cut off the low voltage supply from all outlets in the whole laboratory. The other outlet illustrated is housed in a neat triangular case. It is often more convenient to fix an outlet to the top of the laboratory bench than to recess the outlet in the bench itself and this unit lends itself to such an application. Fig. 21 shows a similar unit containing a mains socket outlet. This can be mounted either independently or side by side with a low voltage outlet.

Connectors

It is often necessary to join cables together. In the wiring of buildings this is never done by soldering. Good soldered joints can be made in factory conditions, but the conditions existing on a building site, and the quality of work that can be done under such conditions, are such that joints would not be sufficiently reliable. Also the time taken to make them would put up the cost of the electrical service considerably. It is, therefore, the practice to join cables by means of connector blocks which require only mechanical terminations to the cables. A connector block is illustrated in Fig. 22. It consists of two screw-down-type terminals solidly connected to each other, mounted in an

Fig. 21   Bench unit (*Courtesy of* The Wandsworth Electrical Mfg. Co. Ltd.)

insulated casing. The end of each wire is pushed into one of the terminals and the screw is tightened on to it. The screw grips the cable, holding it firmly in place and at the same time making a good electrical contact. As the two terminals are solidly connected within the insulated case, the result is that there is a good electrical path between the two cables.

With such connector blocks, it is possible to join cables neatly within the boxes which have already been described. In general, joints should be avoided and single lengths of cable run from one piece of equipment to another, but when an occasion arises when this cannot be done, connector blocks should be used.

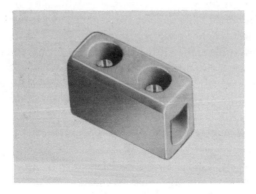

Fig. 22   Connector block (*Courtesy of* Tenby Electrical Accessories Ltd.)

The author has tried in this chapter to give a survey of the more important accessories and to give an idea of the wide range available. It is not possible to describe every accessory made; a full knowledge can be obtained only by a study of many manufacturers' catalogues and, preferably, by the use of the accessories on actual sites.

Hazardous areas

There are industrial processes which involve a risk of fire. Generally the risk arises because flammable vapours are present in the atmosphere. For example in coal mines there is always the possibility of methane appearing in sufficient concentration to ignite or burn. In such cases any electrical equipment in the area subject to risk must be specially designed to reduce the risk.

The mere flow of electricity will not ignite a vapour unless the temperature becomes too high. The temperature can be kept low by adequate sizing of the cables so that this is not a problem. Vapour can however be ignited by a spark at a terminal or switch. The principle adopted in the design of flameproof equipment is therefore that a spark inside equipment should not cause fire outside it. It is not practicable to design equipment so that no air or vapour surrounding it can get inside it. It is however possible to design it so that the air gaps between inside and outside are so narrow and so long that any flame starting inside will be extinguished before it has travelled to the outside. This is the method used for flameproof equipment. BS 229 classifies dangerous vapours into four groups and specifies the dimensions of electrical equipment to be used with each group. Nearly all the accessories described in this chapter, including switches, socket outlets and boxes, are available in flameproof versions complying with these requirements. There are other British Standards giving more detailed information on types of equipment.

Standards relevant to this chapter are:

BS 67	Ceiling roses
BS 196	Protected type non-reversible plugs, socket outlets, cable couplers and appliance couplers
BS 546	Two pole and earthing pin plugs, socket outlets and adaptors
BS 1363	13 amp plugs, socket outlets and boxes
BS 1395	30 amp flameproof plugs and sockets and cable couplers
BS 1778	15 amp plugs, socket outlets and adaptors
BS 3052	Electric shaver supply units
BS 3676	Switches for domestic and similar purposes
BS 4177	Cooker control units
BS 4343	Industrial plugs, socket outlets and couplers
BS 4573	Two pin reversible plugs and shaver socket outlets
BS 5125	50 amp flameproof plugs and sockets

BS 5345   Code of Practice for electrical apparatus in potentially explosive atmospheres.

BS 5419   Air break switches up to and including 1000 V a.c.

I.E.E. Wiring Regulations particularly applicable to this chapter are:

Regulation 411—8
Regulation 471—33
Section 511
Section 512
Section 553

2 Cable

Introduction

Electricity is conveyed in metal conductors, which have to be insulated and which also have to be protected against mechanical damage. The conductors used in domestic and commercial installations are so small that the proper name to give them is wire. When the wire is insulated to make a usable piece of equipment for carrying electricity it becomes a cable. This nomenclature makes a convenient and logical distinction between a bare and an insulated wire, but in practice the terms wire and cable are in fact used interchangeably and it is only the context which makes clear exactly what is being referred to. We shall try to avoid confusion and shall discuss conductors first and the insulation applied to them afterwards.

Conductors

The commonest conductor used in cables is copper. The only other conductor used is aluminium. Copper was the earlier one to be used and although aluminium has no disadvantages, it has to overcome the natural human dislike of change. Its greatest assets are that it is cheaper than copper and that its price is less liable to fluctuations.

Conductors have usually been made by twisting a number of small wires, called strands, together to make one larger wire. A wire made in this way is more flexible than a single wire of the same size and is consequently easier to handle. Each layer is spiralled on the cable in the direction opposite to that of the previous layer; this reduces the possibility that the strands will open under the influence of bending forces when the cable is being installed. Cable made in the United Kingdom in Imperial sizes before the introduction of metric measurements was denoted by the number of strands and the size of individual strands. Thus 7/0.029 meant a wire made from seven strands each of which was 0.029 in. diameter. Larger cables, such as those used for sub-mains, were more usually denoted by their cross sectional area, so that a 0.1 in² cable meant one which had a total conductor cross section of 0.1 in². This method was used for MIMS cables even in the smaller sizes.

When the industry changed to metric units, the manufacturers decided that the smaller cables would not be stranded. Instead, they were made with single wires of sufficient size to carry the rated current. Perhaps because of this, it became the practice that all cables are now denoted by their total cross sectional area. For example, the older engineers who were used to designing lighting circuits to be wired in 3/0.029 cable had to learn to think of 1.5 mm² as the new equivalent size. The latter has one conductor with a diameter of 1.38 mm giving a cross section of 1.5 mm² while the former has three strands each of 0.029 in diameter giving a total cross section of 0.002 in².

A consequence of this change was that the metric cables were not so easy to handle on site. Some contractors got used to this and some complained very loudly. As a result a new range of metric cable has been introduced which is made stranded, but the original single wire type is still in production. At the time this passage is being written, both types are in production but the original single wire type is the more common.

Insulation

Every conductor must be insulated to keep the flow of current within the conductor and prevent its leaving the conductor at random along its length. The following types of insulation are in use.

PVC

Polyvinyl chloride is one of the commonest materials used by man today. It is a man made thermo-plastic which is tough, incombustible and chemically unreactive. Its chief drawback is that it softens at temperatures above about 80°C. It does not deteriorate with age and wiring carried out in PVC insulated cable should not need to be renewed in the way that wiring insulated with most of the older materials had to be. PVC insulated cable consists of wires of the types described above with a continuous layer or sleeve of PVC round them. The only restriction on this type of cable is that it should not be used in ambient temperatures higher than 70°C.

VIR or VRI

These letters stand for Vulcanised India Rubber and Vulcanised Rubber Insulation respectively. The latter is the correct name, but in speech it is almost invariably called VIR, presumably because this way round it comes off the tongue more easily. Wire insulated with VRI was the most commonly used cable before the introduction of PVC.

The copper conductors used in this cable were tinned to protect them from the corrosive action of rubber on copper. The insulation applied to the tinned copper consisted of an inner layer of rubber and an outer coating of vulcanized rubber, with a covering of coloured fabric braiding. This type of insulation is flammable and is not as chemically resistant as PVC. It is especially liable to attack by oil. It absorbs water, which reduces its insulation properties and it becomes brittle with age. Because of the latter characteristic it ultimately fails and installations having this cable must be rewired after 20 to 30 years. When the insulation has become brittle it will stay in place and still perform its function, but the slightest mechanical shock will shake it off the conductor and so bring the likelihood of a short circuit. Hard and brittle insulation is, therefore, one ·of the points to be looked for when an old installation is being examined.

Varnished cambric cables
These do not have the flexibility or toughness of rubber and are susceptible to damp. They are very rarely used, having been almost entirely superseded by PVC insulated cables.

Butyl rubber
This insulation is used for cables which are to be subjected to high temperatures. It is, for example, used for the final connections to immersion heaters, for the control wiring of gas fired warm air heaters and within airing cupboards. It can safely be used for ambient temperatures up to $85°$C. Butyl rubber also has greater resistance to moisture than natural rubber.

Silicone rubber
This is completely resistant to moisture and is suitable for temperatures from $-60°$C to $150°$C. It is undamaged after repeated subjection to boiling water and low pressure steam, and is therefore used on hospital equipment which has to be sterilized.

Although it is destroyed by fire, the ash is non-conductive and will continue to serve as insulation if it can be held in place. A braid or tape of glass-silicone rubber will hold it, and cable made with this construction is very useful for fire alarms.

Glass
Glass fibre has good heat resisting properties and is therefore used for cables which are employed in high temperature surroundings. One example is the internal wiring of electric ovens. Another application which may not at first sight seem to require heat-resisting cable lies in flexible cords for light fittings. Although the object of an electric lamp is to convert electrical energy into light, most of the energy is in fact dissipated as heat. Many light fittings restrict the paths available for the removal of heat and in consequence produce high local temperatures. The high temperature is transmitted to the flexible cord both by

direct conduction through the lamp socket to the conductors and by an increase in the local ambient temperature. If the flexible cord is to last any length of time, it must be capable of withstanding the temperature it is subjected to.

One type of flexible cord is made from tinned copper conductors insulated with two layers of glass fibre. which is impregnated with varnish. A glass fibre braid, also impregnated with varnish, is applied over the primary insulation. This type of cord can be used at temperatures up to $155°$C. If it is made with nickel plated conductors and a silicone based varnish it then becomes suitable for temperatures up to $200°$C.

Asbestos

This insulation is not normally used for building services. It finds application in the wiring of ovens, kilns and similar high temperature equipment.

Paper

Paper insulated cable has been used for power distribution for nearly a century. It is too bulky to be used for the small cables of final circuits within buildings, or for most of the sub-mains. The smallest practicable rating is 100 A, and its chief use is for the Electricity Supply Board's underground low voltage and medium voltage distribution. There are, however, cases in which a large building, such as a school, hotel or hospital requires a large sub-distribution cable from the main intake to a subsidiary control centre, and paper insulated cable is then frequently used.

The conductor is either stranded copper or stranded aluminium, the latter becoming increasingly more popular as its price advantage increases. Whichever is used, it is heavily stranded to give good flexibility, which is important in a cable of such comparatively large size. Paper specially made for the purpose is used as an insulator. It is essential that it should have good mechanical properties to be suitable for this application. Paper itself is a hygroscopic, fibrous material, and has to be impregnated with an oily compound to make it fit for use in cables. The compound used is a heavy mineral oil mixed with resin. On its own impregnated paper insulation would be too fragile to be used unprotected, and a lead sheath is therefore applied over the insulation. Further strengthening and protection can be applied according to the intended use of the cable and the physical wear it may be exposed to. A very good strong protection is afforded by steel wire or tape.

Fig. 23 shows a single core paper insulated lead sheathed PVC covered cable. This is conventionally referred to as a PILSPVC cable. Fig. 24 shows a three core paper insulated lead sheathed steel wire armoured cable with a PVC covering. The abbreviation for this is PILSSWAPVC cable. A considerable number of variations on this basic design is possible and for any given application a cable can only be chosen with the help of a cable manufacturer's catalogue.

PVC is also used for some of the larger power and sub-main cables and may

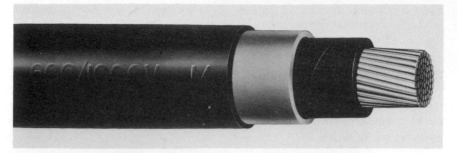

Fig. 23   PILSPVC cable (*Courtesy of* B.I.C.C.)

Fig. 24   PILSSWAPVC cable (*Courtesy of* B.I.C.C.)

Fig. 25   PVCSWAPVC cable (*Courtesy of* B.I.C.C.)

be tending to supersede paper insulated cables for these applications. The construction of such cables is similar to that of paper insulated cables, and an example is shown in Fig. 25. This particular cable would be described as PVC Insulated Steel Wire Armoured PVC Sheathed, which would be abbreviated to PVCSWAPVC Cable. It is rather difficult to find any great differences between the characteristics and behaviour of PVC and of paper insulated cables and the choice is generally made on an engineer's personal preference. The reasons behind personal preferences are not always strictly rational. The example shown has solid aluminium conductors. Identical cables are made with stranded copper conductors.

Air

Air itself is a good insulator. Whilst it cannot of course be wrapped round wires in the ordinary sense of wrapping insulation round, it does form the insulation when bare conductors are used.

Bare conductors are used principally on rising mains in high buildings. These are the mains distributing electric power from the main intake of a building to distribution boards at different levels. The scheme of distribution is discussed in Chapter 5 and here we are concerned only with the construction of the rising main bars. Bare conductors used for rising mains must be correctly spaced from each other to give the necessary air gap for adequate insulation, and should have a protective casing. They must be made inaccessible to unauthorized persons and must have freedom to expand and contract. There are several proprietary systems of bare rising mains, and a typical one is illustrated in Fig. 26. It is frequently used for the vertical distribution in blocks of flats.

The conductors are held in porcelain or sometimes plastic cleats. Apart from supporting the conductors, the cleats keep them the correct distance away from each other for the air gap to have sufficient insulation for the working voltage of the system. The cleats are fixed to the back of a metal trunking which completely encloses the conductors. The dimensions are, of course, such that the air gap between the conductors and the trunking gives enough insulation. The front of the casing is hinged and can be opened for maintenance. It is possible to put a solid insulating plate across the inside of the casing at every floor level to form a barrier to air and smoke moving up and down the casing. In many places this has to be done to satisfy fire prevention regulations. Banks of fuses can be fixed within the casing to form a distribution board as part of the vertical distributor.

A similar system can be used for horizontal distribution. Here again there are several proprietary systems consisting of horizontal conductors supported on insulators inside metal trunking. They are used particularly in factories where they are run horizontally at high level along the walls of workshops. Plain connectors can be fixed at short intervals and short cables run from each set of connectors to a switch fuse fixed on the wall immediately below or above the trunking. The switch fuse can then be connected to serve a machine near it on the floor of the workshop.

MICC

These letters stand for Mineral Insulated Copper Covered; this type of cable is also known as Mineral Insulated Copper Sheathed, which is abbreviated to

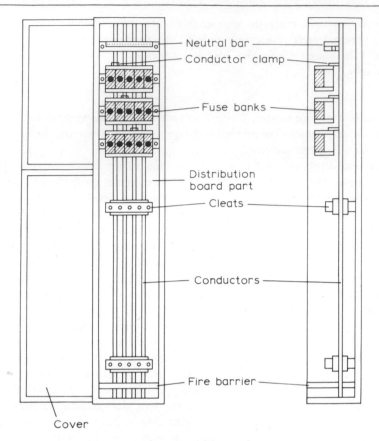

Fig. 26   Rising mains

MICS and as Mineral Insulated Metal Sheathed, abbreviated to MIMS. The last description may refer to aluminium sheathing as well as to copper sheathing. All versions of this type of cable consist of single strand wires embedded in tightly compressed magnesium oxide which is enclosed in a seamless metal sheath. The construction is illustrated in Fig. 27. It is available with the conductors in either copper or aluminium

MIMS cable is extremely robust, and when properly installed has an indefinite life. It can be used outdoors and for such use is usually supplied with an overall covering of PVC. It is then known as MIMS PVC covered. Since PVC is embrittled with ultraviolet light this PVC covered cable should not be installed where it will be exposed to direct sunlight. This is not as drastic a restriction on its use as may appear since it is probably unwise to run any cable where it is so exposed that direct sunlight can get at it. In any such situation, it would also be too vulnerable to damage by vandalism and from animals.

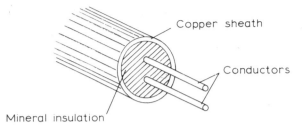

Copper sheath

Conductors

Mineral insulation

Fig. 27   Mineral insulated copper sheathed cable

Because of its robustness, MIMS requires no further protection and is, there-fore, more easily built into the structure of a building than other cables, nearly all of which must have some form of enclosure round them. Because it has an indefinite life, there is no need for facilities to make it possible to rewire the installation. For both these reasons, it can often be used where no other cable would be entirely satisfactory. MIMS cable can carry a higher current than other cables with the same size conductor because the insulation can withstand a higher conductor operating temperature. It follows that for a given current the cable can be smaller if MIMS is used than if another type of cable is used. This is a very useful property which makes it possible to conceal MIMS cable in corners which are not large enough to hide the larger cable that would have to be used with another system.

The magnesium oxide insulation is hygroscopic and will lose its insulating properties if left unprotected against the ingress of moisture from the atmosphere. To prevent this happening, MIMS cable must be terminated in special seals and glands, which are supplied for the purpose by the cable manufacturers. If the cable is cut and the ends left unsealed for any length of time, as can happen in the course of work on building sites, moisture can penetrate the insulation and render the cable useless. In most cases, however, moisture will penetrate unsealed ends for only a short distance of not more than fifty or so millimetres. It is then sufficient to cut off the damaged end of the cable, after which the remainder can be used in the normal way.

Sheathing

We have described how a cable is made from a conductor with insulation round it. Electrically, this is all that is needed to make a device to carry electricity from one place to another, but if the cable is to survive in use, it must also withstand mechanical damage, and the insulation which is enough to achieve electrical protection is seldom strong enough to give adequate mechanical protection. Something further, therefore, has to be provided over the insulation, and it can either be made an integral part of the cable or provided by entirely separate means.

For example, the lead sheath of paper insulated cable, which has been described above, gives mechanical protection and is part of the structure of the cable. Similarly, the metal casing of MIMS cable gives the cable all the mechanical protection that is needed. Neither of these cables could be used without its sheathing, and MIMS could certainly not even be made without it. In these cases, although one may try to draw a logical distinction between the function of insulation and that of sheathing in practice the two must be done together. PVC insulated cable, on the other hand, is strong enough to be handled as it is during erection, but is too liable to mechanical damage to be left unprotected for long. There are two practicable ways of giving it additional protection. One way is to install it in either conduit or trunking and the other way is to put a sheath round the outside of the insulation. The first way gives protection by the use of a particular method of wiring and the second way does it by making the cable a sheathed cable. Methods of wiring are discussed in the next chapter and the rest of this one is devoted to types of sheathed cable.

PVC

Cable insulated with PVC often has a thicker PVC sheath over the insulation and is then described as PVC Insulation PVC Sheathed Cable or simply as PVC/PVC. More than one insulated conductor can be embedded in the same sheath, so that one can have single, twin, three or four core PVC/PVC cables. If one of the conductors is intended as an earth wire, it requires no insulation and may be enclosed directly in the sheath. Fig. 28 shows a cable known as twin and earth PVC/PVC. It has two PVC insulated conductors and one uninsulated conductor all embedded in the same PVC sheath.

Fig. 28   PVC insulated PVC sheathed cable (*Courtesy of* B.I.C.C.)

Rubber

When cable insulated with rubber has a sheath over the insulation, the sheath is also of rubber. VRI cable is given a tough rubber sheath and is hence known as TRS. Except for the difference in material, the construction is the same as that of PVC/PVC cable. Butyl rubber and silicone rubber cables are usually sheathed with thick butyl and silicone rubber respectively.

Flexible cord

Flexible cord is the name given to a particular type of cable. It is one which is flexible and in which the cross sectional area of each conductor does not exceed

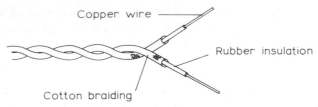

Fig. 29   Lighting flex

4 mm². (In Imperial units the limiting cross sectional area was 0.007 in²). Flexible cords are used for suspending light fittings and for connections to portable domestic appliances having low power consumptions. They are, therefore, usually left exposed in rooms and to be suitable for such use are made with a variety of coverings.

Ordinary lighting flex consists of a stranded tinned copper conductor with rubber insulation, covered with an outer layer of silk or cotton braiding. Two of these cables are twisted together, as shown in Fig. 29 to make a twisted twin flex. Circular flex is made for the connections to household appliances such as irons, kettles and so on, and consists of rubber insulated cable inside an external cotton braiding. Either two or three rubber insulated cables may be used and to obtain the circular cross section cotton padding is wound round them inside the outer braiding. This is illustrated in Fig. 30.

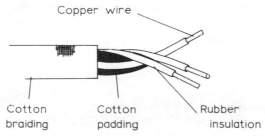

Fig. 30   Circular flex

There are also flexible cords made with silicone rubber insulation with a covering of braided varnished glass fibre. Flexibles which have glass fibre insulation and an outer glass fibre braid have already been mentioned.

Lead
Cables insulated with paper and sheathed with lead have already been described.

Metal
The metal sheathing of MIMS cable is an integral part of the cable and has already been described.

Bare risers
These have already been described. They sometimes have a covering of thin PVC

which is put over them for extra protection to maintenance men who may be working on a system with some of the conductors live. The covering does not perform the protective function of the other kinds of sheathing discussed.

Co-axial cables

Radio and TV systems are now frequently built as part of the engineering services of new buildings. One aerial is made to serve a number of outlets at each of which a receiving set can be plugged in. The single aerial is usually at the highest point of the building or group of buildings in the system and is connected by cable to each outlet. The cable used for this does not carry large currents and is not subjected to large voltages but it does carry a weak signal at high frequencies. The signal must not be lost and to avoid loss of the signal, the cable must have a low impedance at the frequency being used. It must also be constructed so that it does not pick up unwanted high frequency signals, for example, by capacitance or inductance between itself and nearby mains cables.

To satisfy these requirements, radio and TV distribution systems are wired with radio frequency cables. These normally have a single insulated conductor. A metal cover is placed over the insulation to screen the conductor from unwanted signals, and this cover is in its turn protected by an overall sheath of non-conducting material. Thus the single conductor is surrounded by a circular cover and a circular sheath which are concentric and have the conductor on their axis. Hence the cable comes to be known as co-axial cable. Such a cable is shown in Fig. 31; it has its single inner conductor cased in polythene insulation with a wire braid outer conductor and a final sheath of PVC.

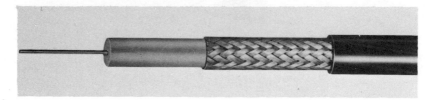

Fig. 31   Co-axial cable (*Courtesy of* B.I.C.C.)

The inner conductor can be either solid or stranded, while the screen can take several forms. It can be a one piece sheath in either aluminium or copper, or it can be of wire braid, which in turn can be either single or double, or it can be of steel tape or of lead. For the insulation, the commonest materials used are polythene, PTFE and polypropylene. The last part of the construction is the outer sheathing and this may be of metal tape, wire, metal braid, PTFE, lead alloy, PVC, nylon or polythene. Evidently, there is a large variety of co-axial cables and properties differ somewhat. The choice of what to use in any particular case is determined by the electrical characteristic required for that particular application and this depends on what equipment the cable is to be used for. Sound broadcasting operates on lower frequencies than television and

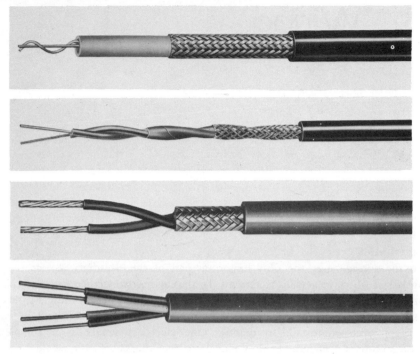

Fig. 32   Audio-frequency cables (*Courtesy of* B.I.C.C.)

cables for sound only do not need to meet quite such stringent conditions as those for television transmission. Audio-frequency cables suitable for microphone and loudspeaker connections for public address systems and for broadcast relay systems are similar to high frequency cables but are not made to such exacting specifications. Examples of audio-frequency cables are shown in Fig. 32.

Standards relevant to this chapter are:

BS 2316   Radio frequency cables
BS 4109   Copper wire for insulated cables and flexible cords
BS 5593   Paper insulated cables
BS 6004   PVC insulated cables
BS 6007   Rubber insulated cables
BS 6195   Insulated flexible cables
BS 6207   Mineral insulated cables
BS 6500   Insulated flexible cords

I.E.E. Wiring Regulations particularly applicable to this section are:

Regulation   412—2
Chapter   52
Appendix   10

3 Wiring

Introduction

To the average user the only important part of the electricity service is the outlets at which he gets his electricity. To the engineer concerned with designing or installing the service, the system of cables which links these outlets to each other and to the supply coming in to the building is just as important and perhaps even more so. In practice, the electrical service is a complete interdependent system and the practical engineer thinks of it as a whole, but, as with the teaching of any subject, one has to break it down into parts in order to explain it in an orderly fashion which will make sense to a student with no previous knowledge of the subject.

In this chapter, we shall consider different ways in which cables can be installed in a building. The calculation of the size of particular cables we shall leave to Chapter 4 and the selection and grouping of outlets to be served by one cable we shall leave to Chapter 5. For this chapter, we assume that we know where cables are to run and discuss only how to get them into the building. This aspect of the electrical service can for convenience be called 'methods of wiring'.

A method of wiring consists of taking a suitable type of cable, giving it adequate protection and putting it into the building in some way. The subject can, therefore, be fairly logically considered by considering types of cable, methods of protection and methods of installation. The types of cable available and in general use have been described in Chapter 2. The protection against mechanical damage given to cable is sometimes part of the cable itself, as with PVC insulated PVC sheathed cables, and sometimes part of the method of installation, as with conduit systems. It can be more confusing than helpful to take a logical scheme of things too rigidly, and rather than deal with protection in a chapter of its own we are dealing with it partly in the previous chapter and partly in this, according to whether it is associated with the cable or with the method of wiring.

It is probably true to say that the commonest method of installing cables is still to push them into conduit and we shall devote most of our attention to this.

Conduit

In a conduit system the cables are drawn into tubing called conduit. The conduit can be steel or plastic. Steel conduit is made in both light gauge and heavy gauge, of which heavy gauge is much more frequently used. In both cases, it can be made either by extrusion or by rolling sheet and welding it along the longitudinal joint. The latter is specified as welded conduit and the former as seamless. Seamless conduit is generally regarded as the better quality. The different sizes of conduit are identified by their nominal bore and in the case of electrical conduit the nominal bore is always the same as the outside diameter of the tube. Thus 20 mm light and heavy gauge conduits both have the same outside diameter and consequently must have slightly different inside diameters. This is the opposite of the convention used for pipes for mechanical engineering in which the nominal bore usually corresponds more closely to the inside than the outside diameter. Electrical conduit is specially annealed so that it may be readily bent or set without breaking, splitting or kinking.

Heavy gauge conduit is normally joined together by screwed fittings; there is a standard electrical thread which is different from other threads of the same nominal diameter. A screwed connection between two lengths of conduit is shown in Fig. 33. A male electrical thread is cut on the ends of both lengths of conduit to be joined and a standard coupler with a female electrical thread is screwed over them. As the two male threads are facing in opposite directions relative to the common female thread, the action of turning the coupler pulls the two ends together. A lock nut, which has been previously threaded well up out of the way on one of the male threads, is then wound down and tightened against the coupler. The screwed connection is relied on for continuity of the earth path and the lock nut is essential to prevent the socket's working its way along the threads until it engages more on one conduit than on the other. The reason for wanting an earth path is discussed in Chapter 9. Methods of joining

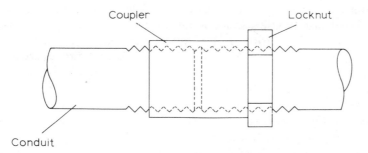

Coupler   Locknut

Conduit

Fig. 33   Conduit coupling

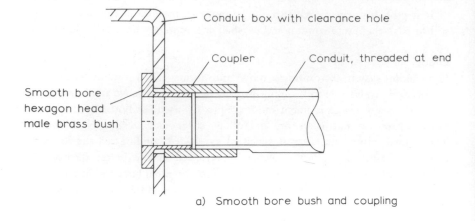

a) Smooth bore bush and coupling

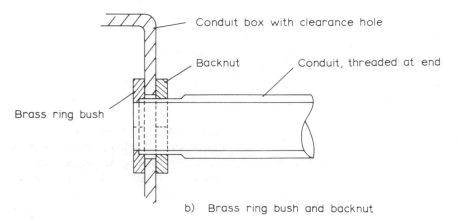

b)   Brass ring bush and backnut

Fig. 34   Conduit entries into boxes

conduit to boxes of the kind described in Chapter 1 are shown in Fig. 34. A bush of some sort must always be used to provide a smooth entry into the box and to avoid sharp corners which could damage the cable insulation. Connections to distribution boards and switchgear are made in a similar manner.

In addition to the boxes described in Chapter 1, other fittings are made for use with conduit. These include the sockets and bushes needed to make connections, and also bends and inspection covers, some of which are illustrated in Fig. 35. The use of bends and inspection covers is not, however, regarded as good practice, because they provide inadequate room for drawing in cable and because they look unsightly when the installation is completed.

Heavy gauge conduit is thick enough for the cross sectional area of the metal to provide a good earth continuity path. The conduit can, therefore, be used as

Fig. 35   Bends and inspection covers (*Courtesy of* Walsall Conduits Ltd.)

the earth conductor and no separate cable or wire need be used for this purpose. When this is done, it is essential that the conduit with all its fittings and screwed joints should form a continuous conducting path of low impedance and the safety of the installation depends on good electrical contact at all the joints. Light gauge conduit is not thick enough to be used as the earth conductor and separate wire must be used. It follows that the conduit connections need not necessarily give electrical continuity and need not be made with quite such particular care.

Light gauge conduit is also too thin to have a thread cut on it. Connections between successive lengths must be made with grip type fittings in which the ends of the conduit are securely clamped by screws tightened down through transverse threaded holes in the coupler.

There is now at least one proprietary system of push on couplings for use with heavy gauge conduit. A picture of it is shown in Fig. 36. The couplers have spring clips which grip the conduit and bite into the metal. It has been shown that these connections are mechanically firm and also that they provide sufficient metal to metal contact to give the necessary electrical continuity, but some engineers are still hesitant to accept them as the equivalent of screwed connections.

Conduit is made in two standard finishes: black enamel and galvanized. It is almost universal practice to use galvanized conduit where it is exposed or where

Fig. 36   Push on coupling (*Courtesy of* Fitter & Poulton Ltd.)

it may be subject to damp. The author of this book used to specify galvanized conduit where there was any chance of corrosion until experience convinced him that black enamel is a better protection than galvanizing and he now thinks that black enamel should be used even when the conduit is exposed to the weather.

The final connection to machines and mechanical equipment such as pumps, boilers, fans, fan heaters, workshop equipment and so on is usually made in flexible conduit. The fixed wiring terminates in a box either in the wall near the equipment to be connected or on the surface of the wall, and from this box a short length of flexible conduit is taken to the equipment. Very often the machine is delivered to the site after the electrician has done the bulk of his work and at the time he is putting in the wiring he does not know where the terminals on the machine are, so that he can only position the outlet box to the nearest foot or so. Solid conduit from this to the machine could involve a large number of bends in a short distance which would be difficult to make and impossible to pull wire through. Flexible conduit can take up a gentle curve and also serves to isolate the fixed wiring from any mechanical vibrations on the connected machine.

There are several types of flexible conduit. Metallic flexible conduit is shown in Fig. 37. It is made from a stepped strip which is wound in a continuous spiral so as to produce a long cylinder with spiral corrugations. The material used is normally galvanized steel. Flexible conduit is also made in a number of plastic materials. In some of these the flexibility is conferred by a corrugated structure, as in the case of metallic flexible conduit, and in others by the flexible properties of the material itself.

Flexible conduit cannot be used to provide an earth continuity path. This is obvious in the case of plastic flexible conduit which is made of non-conducting

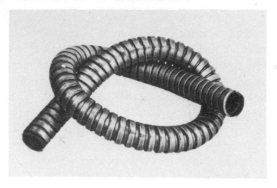

Fig. 37   Flexible conduit (*Courtesy of* Walsall Conduits Ltd.)

material, but it is so even in the case of metal flexible conduit. The flexibility here is conferred by the corrugated structure, and as the conduit bends the corrugations open out. They remain sufficiently overlapping to keep out dirt and moisture but are not in hard enough contact with each other to be relied upon to give an adequate electrical path. To make up for this, a separate earth wire must be run wherever flexible conduit is used. The earth wire is either put inside the conduit with the other cables or it can be placed outside the conduit. In either position, it must be bonded to the rigid conduit at both ends of its run. A clamp for connecting an external earth wire with solid conduit is shown in Fig. 38.

There are other applications for flexible conduit. It is required with certain systems of industrialized building in which sections of floors and walls are precast in factories away from the building site. In order that electrical wiring

Fig. 38   Earth clamp (*Courtesy of* Walsall Conduits Ltd.)

can be put into these slabs after they have been erected, conduit is cast in them and exposed ends are left at the edges where the slabs will be joined together on site. The slabs are lifted into position on the building and joined to each other by *in situ* concrete, grout or some other suitable structural method. At.the same time as this is done, the exposed conduit ends, in adjacent slabs, are linked together by short lengths of flexible conduit. The flexible conduit can take up a gentle 'S' shape and thus make up for some lack of alignment between opposite ends of the fixed conduit. Small errors in casting need not, therefore, cause a problem during assembly, although a very large misalignment will pull the flexible into such a sharp S that the site electrician will not be able to pull cables through it.

A conduit system must be completely installed before any wires are pulled into it. It is, therefore, essential that it is set out so that an electrician can pull cables into it without difficulty. Conduit systems are intended to be rewirable; that is to say the intention is that twenty or thirty years after the building has been erected, it should still be possible to pull all the cables out of the conduit and pull new ones into it. If this is possible, then quite regardless of what happens when the building is first constructed, the layout of the conduit must be such that cables can be drawn into it when it is complete and finished.

The original reason for wanting to have electrical systems which could be rewired during the life of the building was that VRI cable deteriorates in about twenty years to the stage at which it should be removed. PVC cable appears to last indefinitely so that all modern installations which use this cable should not need rewiring. The use of electrical appliances has increased greatly in the last thirty years, and when old buildings which had VRI cable are rewired the opportunity is invariably taken of modernizing the installation by adding extra outlets and circuits. New cables than have to be run where there were no cables previously and the original conduit has at best to be added to and at worst abandoned altogether. Rewirability is then no help and in fact the need for a rewirable system is not as great as is often supposed.

On the other hand, there is always the possibility that a cable may become damaged during the construction of a building, and it is obviously an advantage if it can be replaced without difficulty after the building has been finished. If the conduit is installed so that the system is rewirable repairs will always be possible. The requirements for rewirability should, therefore, be kept to as far as possible, but the engineer in charge should have a discretion to relax them if exceptionally difficult circumstances are encountered.

To achieve rewirability, draw-in boxes must be accessible from the surface, or in other words their covers must be flush with the finished surface. The covers can then be removed without any cutting away of plaster or brickwork. In addition, the length of conduit between successive draw-in boxes should not exceed about 10 m and there should not be more than two right angle bends between successive boxes. A further requirement is that the bends themselves should be made with as large a radius as the position of the conduit within the building permits. This is the reason that specifications often insist that bends shall be formed in the conduit itself and prohibit the use of factory made bends. The latter are necessarily of small radius and could damage insulation if cables have to be forced through them. Inspection bends do not provide adequate room for feeding cables through in a neat and workmanlike manner and the conduit should be so laid out that they are not necessary

Care must be taken in the making of bends to avoid rippling or flattening of the conduit. The smallest sizes of conduit (16 mm and 20 mm) can in fact be bent over one's knee. This is not, however, to be recommended because it is unlikely that a neat bend without kinks will be produced. A bending block, as shown in Fig. 39, is a better device. The bottom edge of each hole should be bevelled so that the conduit is not pulled against a sharp edge. The conduit to be bent is inserted in the hole and hand pressure is brought to bear to bend the conduit slightly. The conduit is then moved through the hole a short distance and the process repeated. Practice is necessary to make a good bend without kinks and not all electricians possess the necessary skill.

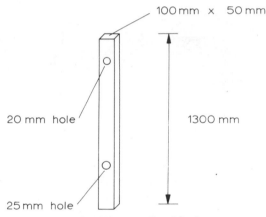

Fig. 39   Bending block

For larger conduit, a bending machine is essential, and is to be recommended for all conduit. It is the only truly reliable way of making a good bend. A bending machine is shown in Fig. 40.

To allow ease of wiring and avoid damage as cables are drawn in, the number of cables in each conduit has to be limited. The I.E.E. Regulations give tables from which one may determine the size of conduit required to carry various numbers of cables. The tables apply to the commonest situations and for other cases the Regulations stipulate that the space factor shall not exceed 45%. The space factor is defined as

$$\frac{\text{total cross sectional area of cables}}{\text{internal cross sectional area of conduit}} \times 100\%$$

Fig. 40   Conduit bending machine (*Courtesy of* Record Ridgway Ltd.)

It is harder to pull several small cables together than one large cable, and when a number of cables have to go in the same conduit, it is advisable to keep the space factor well below 45%. Space factors of less than 20% need not be considered at all extravagant. For the same reason, it is often better to use two size 25 mm conduits side by side than a single 32 mm or 50 mm even when in theory the latter is adequate.

Many types of insulation deteriorate if they become damp. It is, therefore, important that moisture should not collect in the conduit system. Moisture can occur through water getting in during building operations and also later on through condensation of moisture in the atmosphere. A conduit system must be laid out so that it is well ventilated, which will prevent condensation, and so that water which does get in will drain to one or more low points at which it can be emptied.

It is good practice to swab through the conduit after it is erected and before cables are drawn in to remove any moisture and dirt which have collected. This is simply done by tying a suitable size of swab on the end of draw wire and pulling it through the conduit from one draw-in box to the next.

To avoid damage to cables as they are drawn in, burrs on cut ends of conduit must be removed with a reamer before the lengths of conduit are joined.

There are a number of positions in a building in which the conduit can be fixed. It can obviously be run on the surface of walls and ceilings, and when a building is constructed of fair faced brick walls, surface conduit is usually the only practicable wiring system which can be adopted. If walls are plastered, the conduit can generally be concealed within the plaster. There must be at least 6 mm of plaster covering the conduit if the plaster is not to crack. Since plaster depth conduit boxes are 16 mm deep, the total thickness of plaster must be at least 22 mm. If the architect or builder proposes to use a lesser thickness than this, it becomes necessary to chase the conduit into the wall so that some of the total distance of 22 mm between face of plaster and back of conduit is in the wall and some in the plaster. This is shown in Fig. 41.

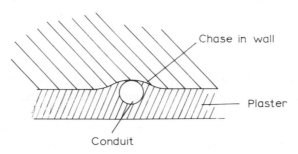

Fig. 41 Conduit chased into walls

In many modern buildings, internal partitions which do not carry any of the structural load are made of breeze blocks about 75 mm thick and in some cases as little as 50 mm thick. If these have to be chased to take 25 mm conduit, there is very little partition left. Using conduit with such partitions is a very real problem and the electrical engineer often has to abandon a conduit system in favour of one which is less robust but takes up less space.

Horizontal runs of conduit over floors can sometimes be arranged within the floor finish. Probably the most widely used floor finish is still the fine concrete screed. Provided the screed is sufficiently deep, the conduit is laid on top of the floor and just screeded over. If the screed is not deep enough to make this possible, it may be possible to cut chases in the floor itself so that the conduit is partly in the floor and partly in the screed. However, structural floors are nowadays designed to such close limits that the structural engineer may not permit the electrical engineer to have chases cut in the floor slabs. It is often necessary for conduits to cross each other in a floor and there are also other services, such as water and gas, which run in pipes laid in or over the floors. It is then almost inevitable that conduit has to cross one or other of these other services. It will be obvious that crossovers, whether of conduit and conduit, or of conduit and other services, are the places at which maximum depth is needed. It is these critical points which determine whether or not it is possible to accommodate the conduit within the floor finish. This must be discussed by the electrical engineer and the architect quite early in the design as the decision will affect the type of wiring which the engineer has to design.

Conduit can also be buried within concrete slabs. Many modern buildings have floors and even walls of concrete with very little or no finish on top of it; this is particularly true of industrialized methods of building. In such buildings, the only practicable alternative to putting wiring on the surface is to bury conduit within the structural concrete. This needs considerable care. The exact position of the conduit within the depth of the slab must be agreed with the structural engineer and close supervision is required of the work on site to ensure that the conduit is correctly placed. It has to be fixed in position immediately after the steel reinforcement has been laid in the shuttering and before the concrete is poured. If it is not well tied either to the reinforcement or to the shuttering, it may be dislodged as the concrete is poured and vibrated. Open ends of conduit which may have to be left at the end of the section of concrete being cast, ready for connection to the next piece of conduit, must be covered with metal or plastic caps to prevent cement or stones getting into the conduit. Every electrician of any experience can tell his horror story of blocked conduit. The conduit boxes must also be filled with a material which will prevent cement and stones getting in but can itself be easily removed once the concrete has set and the shuttering has been struck. The most commonly used material for this purpose is expanded polystyrene.

Once conduit has been cast inside a concrete slab, it is totally inaccessible for repair or replacement. The rules for installing it in such a way that the drawing in of cables is easy are, therefore, of exceptional importance. It is advisable for the conduit to have plenty of spare capacity for the number of cables to be drawn into it, for bends to be very easy and for there to be plenty of draw-in boxes.

Lightweight conduit is unsuitable for burying in the building structure. It is difficult to make the joints watertight, and because of its thinness it is liable to be disturbed and damaged when concrete is poured over it. When conduit is placed on top of a floor ready to be screeded over, workmen are liable to walk over it after it is laid and before the screed is poured. Light gauge conduit is not robust enough to stand up to this. The use of lightweight conduit is, therefore, usually confined to small domestic installations.

In wooden floors, conduit can be run under the floorboards. Where it has to run across joists, the latter must be slotted for the conduit to get through underneath the floorboards. The agreement of the structural designer must be obtained before joists are cut. This method is not, however, much used now; in wooden floors it is more usual to employ PVC sheathed cable clipped to the joists. It does not need further protection.

Some thought has to be given to the relative position of conduit and boxes. The position of the conduit is determined by the route it takes through the structure. The outside of the box has to be flush with the finished surface, or in the case of a surface system the back of the box must be on the surface. The positions of the conduit and box being fixed independently of each other by different considerations, it may happen that the conduit is not in line with any of the outlet holes in the box, and some method has to be devised to overcome this mismatch.

Fig. 42a shows surface conduit with a set in it to enter a surface box. It is difficult to make this look neat and it is better to use a distance saddle and a special box which makes it unnecessary to set the conduit. Such a box is shown at Fig. 42b. When the conduit is buried in the structure, it may have to be set as shown in Fig. 42c. If the conduit is far enough inside the surface, a back entry box can be used as in Fig. 42d, but it must be remembered that this introduces a fairly sharp bend in the conduit which could make it harder to pull the cable in. Another possibility is to put the box in line with the conduit and fit an extension ring to the box to bring the cover forward to the surface. This is shown in Fig. 42e.

When buried conduit has to feed surface distribution boards or switches, the conduit must be brought into a flush recessed box so that the cables enter the surface board or switch through the back. If necessary an extension ring has to be placed between the box and the surface. Fig. 42f shows an example of buried conduit feeding a fuseboard on the surface.

Most buildings larger than a single dwelling have a three phase supply, although nearly all the equipment in them is single phase. Circuits on different phases should not be taken through the same conduit. Regulations issued by the Institution of Electrical Engineers also make it necessary to separate normal voltage circuits (i.e. 240 V, but includes any voltages down to 50) from low voltage and telecommunications circuits (i.e. 50 V or less) and from fire alarm circuits, and to keep each of these three categories in conduit separate from that holding either of the other two, but beyond the requirements of regulations it is good practice to keep different circuits within each category away from each other. Thus, although it is permissible to run lighting and power circuits in the same conduit, it is good practice to separate them. Similarly, although the regulations allow one to run television and telephone wires in the same conduit, it is better not to do so.

The next matter to receive our attention is how to fix conduit. Fig. 43 shows various devices for fixing conduits. The pipehook or crampet, Fig. 43a, is a satisfactory and simple fixing, but is too unsightly to be used on surface work. It can be driven into timber, brick or masonry, but is more likely to be dislodged

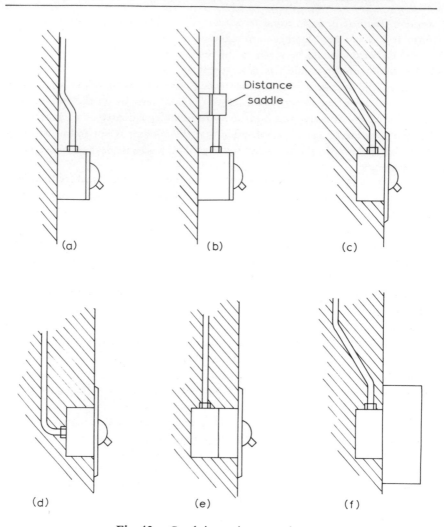

Fig. 42   Conduit entries to equipment

than a screwed fixing; where the conduit is to be buried in plaster after it has
been fixed, this does not matter because the plaster will hold the conduit in
place, but where the conduit is to remain exposed a firmer fixing is desirable.
The saddle hook shown in Fig. 43b is by far the commonest fixing. It passes
round the conduit and is secured to the wall by two screws. The only advantage
of the clip shown in Fig. 43c is that it saves one screw. It is not as secure as the
saddle and the cost saving is not sufficient for a good engineer to use it.

Sockets and other conduit fittings necessarily have a larger outside diameter
than the conduit itself. If these components are tight to a wall, the conduit must

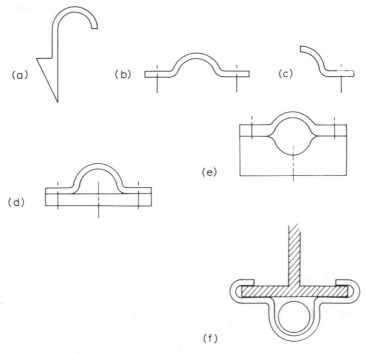

Fig. 43   Conduit fixings

be slightly proud of the wall. Because of this, when an ordinary saddle is tightened, it will tend to distort the conduit. This can be prevented by the use of a spacer saddle, Fig. 43d, which has the same thickness as the sockets. The spacer saddle has the further advantage that it prevents the conduit from touching damp plaster and cement which could corrode and discolour decoration.

When conduit is fixed to concrete, the time taken to drill and plug holes in the concrete is a very large proportion of the installation time. A spacer bar saddle has only one screw to be fixed to the wall and the saving in time can be greater than the extra cost of the material.

The distance saddle shown in Fig. 43e holds the conduit about 10 mm from the wall. This spacing eliminates the ledge between the conduit and the wall where dust can collect and makes it possible to decorate the wall behind the conduit. It also makes it impossible for minute drops of moisture to collect in the crack between conduit and wall and thus reduces the possibilities of corrosion. For these reasons, distance saddles are almost invariably specified by hospital boards and local authorities for surface conduit.

Conduit often runs across or along steel girders or joists, either exposed or within a false ceiling. It is not desirable to drill and tap structural steelwork and it is better to use girder clips of the type illustrated in Fig. 43f. Whilst standard girder clips can be bought from conduit manufacturers, it is usually simpler to make special clips to suit individual conditions on each job.

Electrical conduit is not thick enough to support its own weight over long distances without sagging. The supports must, therefore, be at quite close intervals, and the maximum distances which should be allowed between supports are as follows:

20 mm conduit   1.5 m
25 mm conduit   2.0 m
32 mm and over   2.4 m

The cables are drawn into the conduit with the help of a steel tape and a draw wire. The steel tape has a hemispherical brass cap on the end which prevents its sticking on sharp edges and irregularities at joints of the conduit and also helps guide it round bends. The tape also has a loop at its other end and a steel draw wire is attached to this. The cables themselves are then attached to the other end of the draw wire. The electrician attaches the cables by threading them through a series of loops in the draw wire. They should not all be attached to the same point otherwise there is a large sudden enlargement in the bunch of cables and this presents an edge which can catch in the bore of the conduit and which will be difficult to get round bends. When each cable is looped through the draw wire, it is folded back on itself and the end is taped. This gives a smooth surface to be pulled through the conduit and prevents sharp ends of small strands of wire from sticking out and catching the inside of the conduit. The method of connection is shown in Fig. 44.

Pulling cable through conduit is a job for electrician and mate. One man pushes the steel tape with the draw wire attached to it from one draw-in box to the next. As soon as it appears at the receiving box, the second man takes it and pulls gently from that end. This man then pulls the draw wire and finally the

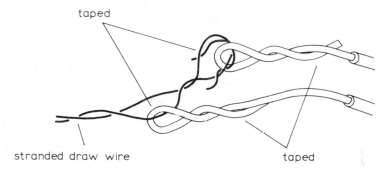

taped

stranded draw wire   taped

taping over loops and ends of wires not drawn

Fig. 44   Connection of cable to draw wire

bunch of cables while the first man feeds them into the conduit. The man feeding the cables in must do so carefully and must guide the cables so that they do not cross or twist over each other as they enter the conduit. If they are allowed to twist, the whole bunch may stick and even if they can be forced in, it may be impossible to withdraw some of them later. The whole job requires great care and needs co-operation between the two men at opposite ends of the run. It is a help if they are within sight of each other and essential that they should be within earshot of each other. On the rare occasions when the run of conduit through the building from one draw-in box to the next makes it impossible for shouted directions to be heard from one end to the other, a third man will have to be called in to stand halfway and relay messages.

In hot weather, the insulation of the cables is liable to become soft and tacky. Drawing it through the conduit can be made easier by rubbing French chalk on the cables. In other circumstances, when friction between the cables and the conduit is high and makes pulling in difficult, it may be advantageous to apply a thin coating of grease or tallow wax to the cables.

Plastic conduit

PVC conduit is being increasingly used in place of heavy gauge steel conduit. Its advantages are that it is cheaper and more easily installed than steel conduit and that it is non-corrosive and unreactive with nearly all chemicals. Although it is incombustible, it does soften and melt in fires and cannot be used at temperatures above 65°C. At low temperatures, it becomes brittle and should not be used where it will be exposed to temperatures below 15°C. Most specifications call for high impact grade heavy gauge PVC, which is tough enough to withstand the ill treatment which all material gets on building sites. It will protect cables inside it from nails accidentally driven into the conduit just as well as steel conduit can, but it is not quite as resistant as steel conduit to heavy blows and to crushing.

Heavy gauge PVC conduit is not resistant to blows, but has a slightly higher temperature range than the high impact grade, and is suitable for many types of industrial installation. Light gauge conduit is cheaper but not so robust and may not always withstand the conditions existing on a building site.

As the conduit is made of an insulating material, it does not provide a means of earth continuity. A separate earth wire must, therefore, be pulled into the conduit along with the other cables. A plain bare wire of adequate cross section may be used for this purpose, but the hard metal is liable to damage the insulation of the other cables as they are being drawn in. It is better to use a PVC insulated cable for the earth wire as well as for the phase conductors. When a separate earth wire is used, it may be necessary to connect lengths of it at wiring points. Some manufacturers, therefore, provide PVC conduit and switch boxes with an extra earth terminal. In most cases the boxes can be ordered with or without the extra terminal.

As explained in the description of steel conduit, there are situations in which flexible conduit has to be used. In a steel conduit system, the flexibles do not provide earth continuity and a separate earth wire has to be run along each flexible length. If a large number of such connections occurs, one of the chief advantages of a steel conduit system, namely the way it gives earth continuity, is lost. In that case, it may be as well to use PVC conduit throughout with a separate earth wire throughout.

Lengths of PVC conduit are joined by an unscrewed coupler which is cemented to the two pieces of conduit to be connected by means of a special solvent. The solvent used is made particularly for this application and is supplied by the makers of the conduit. PVC conduit boxes for use with PVC conduit have short sockets which make it possible to connect the conduit to the box with a coupler of the same type as is used for connecting lengths of conduit. The PVC can also be threaded, and push fit to threaded adaptors are made with the aid of which connections to boxes and equipment can be made in the same way as for steel conduit.

PVC has a high coefficient of expansion and provision must be made for thermal expansion wherever there is liable to be a temperature change of 25°C or more and also wherever a run of more than 8 m occurs. The necessary allowance is made by means of expansion couplers. An expansion coupler is a coupling of extended length, one end of which is bored to a standard depth and the other end of which has sliding fit over a longer distance than the standard coupling depth. Expansion is liable to make PVC conduit sag more readily than steel conduit and it needs fixing at closer intervals. Saddles should be fitted at a spacing of about 900 mm

Bends can be made in PVC conduit as in steel conduit, but it is essential to use a bending spring inside the conduit to prevent the cross sections becoming reduced in the bending process. The smaller sizes can generally be bent cold, but 32 mm conduit and larger must be gently heated for a distance of about 300 mm on either side of the intended bend.

One has to remember the susceptibility of PVC to high temperatures if one proposes to suspend light fittings from PVC conduit boxes. The heat from the lamp is conducted through the flexible cable and through the fixing screws, and it could happen that it softens the PVC box. To overcome this problem, it is possible to provide the boxes with metal inserts.

PVC conduit is made not only in the normal circular cross section but also with an oval section. The reduced depth of an oval section enables it to be accommodated within the thickness of plaster in places where the use of round conduit would make it necessary to chase the brickwork behind the plaster. This makes the oval conduit very useful for switch drops and for small domestic installations. In the latter case, it makes it very easy to add new wiring in an old house. The electrician can cut away and repair plaster whereas he would probably want help from another tradesman if he had to cut into brickwork.

The same PVC material is made as rectangular and semicircular channelling. This is intended primarily as a protection over PVC insulated PVC sheathed cable where the latter is installed on the surface of walls. It can also be used as a protection to PVC/PVC cable when the latter is buried in plaster, the justification for this use being that it saves depth and that the side of the cable next to the structural part of the wall does not need protection.

Flexible PVC conduit is available in two types. In the one, flexibility is conferred by a corrugated construction. In the other, the PVC itself is a plasticized grade so that the flexibility is a property of the material itself. Flexible PVC conduit can be used to negotiate awkward bends and in situations where rigid conduit would be difficult to install, and it is sometimes resorted to for the solution of unforeseen problems which so often seem to arise in the course of building work. There is, however, a danger to using it in this way. It is possible to take such advantage of the flexibility that the conduit curves so sharply that it is impossible to pull cables through it. If this happens the problem of installing the conduit has been solved only by the creation of a more difficult problem for the next stage of the erection process. Flexible conduit should, therefore, be used with caution.

PVC sheathed cable

There are many cases in which wiring can be installed in PVC/PVC without further protection. For example, this may be done in any voids in a building such as false ceilings and wooden floors. When the cable runs parallel to joists in a wooden floor, it can be clipped to the sides of the joists. When it has to run across them it is better to thread it through holes drilled in the neutral axis of the joists than to notch the top of the joists. Holes drilled on the neutral axis weaken the joist less than notches cut in the top, and because the cable is further from the floor boards on top of the joists, it is safer from nails driven into the floor. New buildings often have internal partitions constructed of timber studding with light plasterboard facing, as illustrated in Fig. 45. This is particularly the case with proprietary industrialized systems of building. Such a partition has a void within it which is used for engineering services, and this is also a situation in which PVC sheathed cable without further protection is the most suitable system of wiring. Voids of this sort in which PVC sheathed cable is run are usually sufficiently accessible to make rewiring, if not easy, at least possible.

It may happen that the building structure is such that PVC sheathed cable on its own is the most suitable system to use, but that there are a few places where cable has to drop in plastered walls or run across floors. It is then desirable to give the cable additional protection at these places by running it inside conduit at these places only. This has the additional advantage of making rewiring easier. As the conduit is used only for short lengths for local protection both light gauge steel and PVC conduit are suitable.

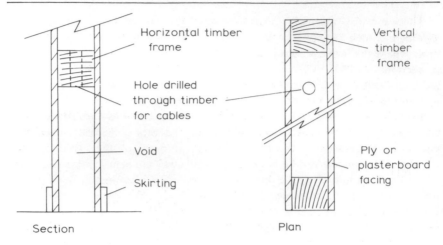

Fig. 45   Dry partition

PVC sheathed cable may be buried in plaster without damage. There is, however, the possibility that nails may be accidentally driven into the cables when pictures are being fixed to walls. Ideally, the cable should therefore be protected by conduit, but if there is not sufficient depth of plaster to make this possible it can still be given protection by shallow rigid PVC or galvanized metal channelling as shown in Fig. 46. Many authorities feel that even this is not necessary. The author has been rather cautious about specifying PVC sheathed cable in plaster without protection, but after discussion with many colleagues has accepted that there is sufficient experience, both in the United Kingdom and on the Continent to show that the probability of an accident is low enough to make unprotected PVC sheathed cable in plaster an acceptable system.

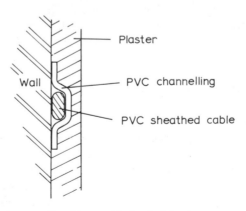

Fig. 46   Cable buried in plaster

There are situations where the appearance of an installation is of secondary importance, and where at the same time a surface system will not receive rough usage. Such a case might occur in an old building used for commercial purposes or in simple huts at a holiday camp. PVC sheathed cable may then be run on exposed surfaces without further protection. Since it is visible it will not be damaged accidentally by people trying to fix things to the walls.

PVC sheathed cable can be fixed either with moulded plastic clips or with buckle clips. These are illustrated in Fig. 47. The clips should be spaced not more than 200 mm apart.

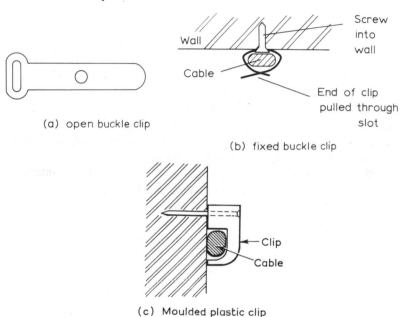

(a) open buckle clip

(b) fixed buckle clip

(c) Moulded plastic clip

Fig. 47   Clips

TRS cables

These are no longer in common use, but could be installed in exactly the same way as PVC sheathed cables.

Cable trunking

Where a large number of cables has to be run together, it is often convenient to put them in trunking. Trunking for electrical purposes is made of 18 gauge sheet steel, and is available in sizes ranging from 50 mm x 50 mm to 600 mm x 150 mm, common sizes being 50 mm x 50 mm, 75 mm x 100 mm, 150 mm x 75 mm and 150 mm x 150 mm although 50 mm x 100 mm and 100 mm x 100 mm are also available. It is usually supplied in 2 m lengths and one complete side is removable, as shown in Fig. 48. The removable side, or lid, either screws on or clips on with a snap action. The latter arrangement is cheaper but a little more awkward to handle.

A variety of bends, tees and junctions is available from all manufacturers of such trunking. Some of these are shown in Fig. 48. They enable the trunking to be taken round corners, to reduce in size as the number of cables is reduced and to allow a main run to serve a number of branches.

To put cables in such trunking one normally takes the lid off, lays the cables in and replaces the lid, but it is possible over short distances or straight lengths to pull the cables in as one does with conduit. Whichever method is adopted the number of cables and size of trunking must be such that a space factor of 45% is not exceeded.

Being so much larger than conduit, trunking can quite clearly not be buried in the walls of a building. It has to be run on a surface. There are occasions when there are many circuits running together inside a builder's work vertical duct which also contains other services such as heating pipes or gas. This is one situation in which cable trunking is an ideal way of installing the wiring. It can be similarly used in false ceilings. In both these cases, there must be sufficient doors or access traps to enable electricians to get to the trunking for rewiring.

Buildings such as assembly halls and gyms often have exposed steel lattice framework supporting the roof. It is then possible to run cable trunking neatly through the spaces of the lattice. Sometimes the architect will permit the cable trunking to be fixed under the beams and along them. Either method is simpler

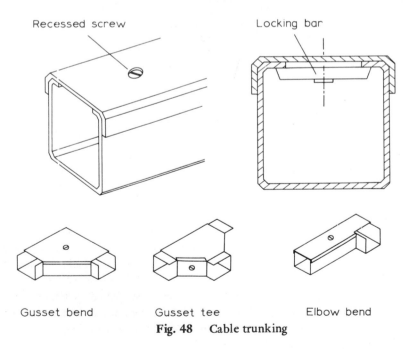

Fig. 48   Cable trunking

and neater than fixing several conduits parallel to each other on the surface of the ceiling, and has the further advantage that during the life of the building, the wiring can be altered very easily.

In workshops and laboratories there is usually a large number of machines and other equipment which have to be served with electricity. A conduit system can then become complex and, therefore, expensive. A simple and neat method of wiring these areas is to run trunking round the walls and to install all the circuits inside the trunking. In rooms of this class, this is quite acceptable and no one objects to the appearance of trunking visible on walls. Again there is the advantage that when machines are replaced or when new machines are installed, the consequent changes to the electrical service are easily made. The same consideration applies, but with added force, to factories.

When machines are placed in the centre of a room a good method of serving them is to run trunking at high level under the ceiling and drop to each machine with a length of conduit. It is, of course, possible to install conduit within the floor with an outlet near each machine, but there then has to be either rigid or flexible conduit at floor level, and if machines are moved or additional ones brought in the floor has to be dug up before the conduit can be extended to the new positions. For the initial installation the electrician would have to know the exact positions of the terminals of each machine before the floor is laid, and it is very seldom that either the builder or the final occupier of the factory can give him this information so early. The overhead system avoids the difficulty of locating exact positions of machines too early in the construction process and makes future changes more easy.

In some cases, particularly woodwork rooms in schools and colleges, long pieces of material such as timber have to be carried from stores across the room to various machines. It can then happen that vertical drops of conduit from the ceiling to the machines obstruct the material and cause difficulty in handling it. If this is likely to happen there may be no alternative to installing conduit within the floor, but the customer's attention should be drawn to the inflexibility of this arrangement.

It is also possible to install cable trunking neatly in the corner between a wall and ceiling. This is sometimes acceptable to an architect, and the author knows of at least one case where it has been done along the length of a public corridor in a block of flats. Similar situations could arise in offices and hospitals. It may be necessary to run many circuits the length of a corridor in such buildings and a conduit system would require many conduits cast into the ceiling slab of the corridor. If the slab is thin and heavily loaded structurally, as is apt to happen in modern building design, it becomes quite a serious problem to get the conduit into the slab within the restricted width of a corridor. The difficulty is avoided if the architect can be persuaded to accept a neat piece of surface trunking. There are now several proprietary systems of trunking specifically designed for such an application and giving a reasonably neat finished appearance.

The sides of cable trunking can be drilled to make holes at which conduit can enter the trunking. Joining conduit to trunking may be convenient where one circuit leaves a line followed by several circuits. All such holes must be fitted with smooth bore bushes to avoid sharp edges which could damage insulation. If PVC sheathed cables run for part of their length in trunking, that is to say continue free without further protection, the edge of the hole in the trunking must be protected by a rubber grommet.

Plastic trunking

Cable trunking is also made out of rigid high impact PVC. In this case, the lid clips on, as will be clear from the cross section in Fig. 49. Bends and fitting similar to those for steel trunking are made, and the plastic trunking is used in exactly the same way as steel trunking.

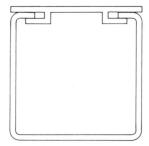

Fig. 49   PVC trunking

Proprietary systems

A number of proprietary systems of wiring has been introduced in the last few years. Some of these are intended specifically for use with particular methods of industrialized building, whilst others are intended to industrialize the wiring operation independently of the building method so that they should be equally useful for all methods of building. One of the latter is the Simplex PVC System.

This consists of ordinary PVC insulated cables drawn into a PVC conduit of oval section. The earth wire is moulded in with the conduit, as can be seen from Fig. 50. Each of the other cables has its own passageway within the conduit. The conduit itself is made from a flexible grade of PVC and the entire assembly of insulated cable within conduit is flexible. From the installer's drawings, the manufacturers make a kit consisting of cable already within the conduit, cut and formed to the lengths needed on the building. The kit is delivered to site and simply laid in position, a process which eliminates most of the ordinary site work required for a conduit system. Although it is supplied with the cable already in the conduit, the system is rewirable, because each cable is loose in its own passage. Suitable conduit boxes for junctions and accessories are supplied as part of the system.

There are problems which can arise with such a system. The conduit and cable are cut to size at the factory from drawings. A slight error in setting out of the building can result in the kits not fitting at site, with many unpleasant recriminations between designer, supplier and installer. Discrepancies can occur even if dimensions are taken on site; usually a typical house or room is measured and all of that type are made the same. If the building setting out varies from one to another, some of the kits may not fit exactly and the site electrician may have some adjustments to do. If every room has to be individually measured, a great deal of the time saving of the system is lost.

Another system takes the form of trunking in the shape of skirting, which is placed round the bottom of walls in place of ordinary skirting. The system includes suitable corner pieces and boxes to hold socket outlets and other accessories. This particular system is very useful for buildings constructed from concrete slabs precast in moulds, but it is still necessary to find some way of running switch drops in the wall and of taking cables from one room to another without blocking door openings.

There are several other proprietary systems, which generally tend to be variations on these themes, and which all have their individual advantages and disadvantages.

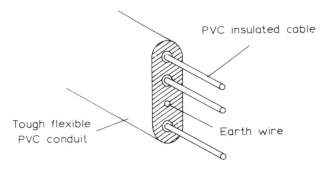

PVC insulated cable

Tough flexible
PVC conduit

Earth wire

Fig. 50   Simplex system

MICC

Mineral insulated cable has been described in Chapter 2. It was there explained that its chief advantage is that it needs no protection and can be put into places where it would be difficult to install other cabling systems. The fact that it needs no protection and is so robust makes it very easy to install.

It can be clipped to walls and ceilings in a similar way to PVC sheathed cable. Sharp bends should, however, be avoided, and a safe rule is to keep the radius of each bend to more than six times the cable outside diameter. Clips or saddles should be fixed at a spacing equal to 75 times the cable outside diameter. If the cable is to be buried within the structure, it should be fixed down firmly before concrete or cement is poured over it.

Accessories are contained in standard boxes and the MICC cable is brought into the boxes. Where it enters a conduit box or the terminal box of a machine, it ends in a seal and a gland. The seal forms the end of the cable and prevents moisture getting into the mineral insulation; it seals the cable. The gland joins the cable to the cable entry on the box, switch or other equipment and provides earth continuity between the cable sheath and the box or equipment case. An assembled seal is shown in Fig. 51 and the procedure for making it is as follows.

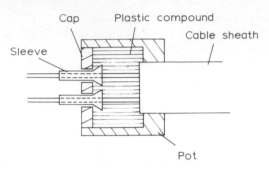

Fig. 51   MICC cable seal

The end of the cable is cut to length and a notch is made round the sheath with a ringing tool. The sheath is then stripped off the cable, from the cable end to the notch, and the mineral insulation is broken away to expose the conductors. At this stage, the gland should be slipped over the cable and pushed up out of the way while the seal is being finished. The gland will then be in place to be brought forward over the seal after the latter is completed.

The cap and insulating sleeves are now assembled. The sleeves are cut to the length required and a wedge is inserted into the end of each sleeve to expand the end of it. When the sleeve is pushed through the hole in the cap, the wedge prevents it from being pulled right through and keeps it in place after the seal is assembled.

The next step is to push the pot over the end of the cable. It is made to suit the cable and is a fairly tight fit, so that it has to be screwed on. It is screwed on until the cable sheath is level with the shoulder at the base of the pot. The pot is then packed with a plastic compound which is pressed firmly in to fill the whole of the pot. The cap and sleeve sub-assembly is now pushed over the ends of the wires and forced into the end of the pot, which is crimped over the cap. This final operation can conveniently be done by a combined compression and crimping tool specially made for this particular job.

The gland is shown in Fig. 52. It consists of a gland body, a compression ring and a gland nut. The compression ring is comparatively soft and as the nut is tightened onto the body the ring is compressed between them and deforms. As it does so, the sharp edge on its inside diameter bites into the outside of the cable

sheath and makes a firm electrical and mechanical bond between the sheath and
the gland. The gland is thus fixed to the cable over the seal and has a projecting
male thread which can be inserted into a fitting or conduit box and secured by a
ring bush and back nut.

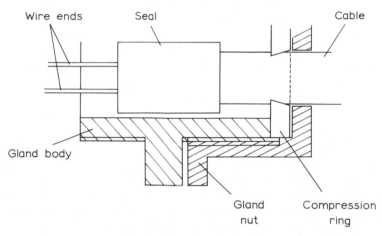

Fig. 52   MICC gland

External wiring

It is sometimes necessary to run cables from one building to another. This can be
done either with cables running under the ground or with overhead cables.
Cables that are to be run underground can be armoured paper, armoured PVC or
MICC cables. All these are strong enough to be laid directly in the ground and
buried, but in order that they should have reasonable protection against damage
they should be laid at least 600 mm below the ground level. Since they all have
metal armouring or sheathing no protection in the ground is really necessary, but
it is very usual to cover the cable loosely with tiles before the trench in which it
has been laid is backfilled. If a workman later has occassion to dig the ground
near the cable he will hit the tiles first and they will warn him that there is
something underneath them. Tiles are available for this use which have lettering
on them saying 'Danger — Electric Cables'; they are known as electrical tiles.

It is also possible to bury conduit in the ground and pull cables through it in
the ordinary way. Because of the danger of corrosion in the soil this is not
however a good practice. It is better to use builder's earthenware ducts instead
of conduit, and this is in fact very often done. It is harder to provide frequent
access to underground ducts than to conduit in a building, and the lengths
between draw-in points can become rather large. At the same time there is
plenty of room in the ground for large diameter pipes, and it is a sound
precaution never to use anything smaller than a 100 mm earthenware duct.

In certain cases it is common practice to use polythene conduit underground.

This happens for example when a communal TV aerial has to cross from one building to another on the same site. It is also the standard method of bringing telephone wires into a building in urban areas where the main telephone cables are in the road outside the new building. Telephone wires are quite small and can be easily pulled into 20 mm conduit over considerable distances.

Buried cables rely on conduction of heat through the soil to dissipate the heat generated by the current in the resistance of the wires. If there are other services which heat the soil locally then the rate of dissipation of heat could be reduced, and the current carrying capacity of the cables would then also be reduced. The obvious example of a service which would have this effect is a district heating main. But in addition to this consideration some thought must be given to what happens when maintenance work is done on underground services. It is undesirable that workmen who may have to expose a length of buried gas or water main should have to dig near a live electric cable. There are thus two reasons why underground cables should be kept well away from other buried services. A good practical rule is to have a minimum distance apart of 2 m.

There may be cases where cables have to cross from one building to another overhead. This situation will normally arise where cost is the overriding factor and neither restrictions on headroom nor appearance are of great concern. For very short distances, as for example from a house to a shed, a piece of conduit can support is own weight over the gap to be spanned and an ordinary conduit system can be used. This method is not suitable for distances of more than about 10 m. For larger distances special overhead cables must be used. Cables made for this purpose consist of solid drawn copper or aluminium conductors covered with a PVC sheath and are suitable for spans of up to 30 m. They can traverse greater distances provided they are supported every 30 m. If supports cannot be found on walls or roofs then wooden or metal poles must be erected to carry cleats on which the cable can be supported.

The limitation on span can be overcome by the use of a catenary wire, which can be used in one of two ways. A separate catenary wire of adequate strength can be strung between the two end supports and a sheathed cable can be suspended from the catenary at regular intervals of about 2 or 3 m.

Alternatively it is possible to obtain cable which incorporates a catenary wire within the sheath. Such cable is specially made for overhead use, and the manufacturer's recommendations on spacing should be noted and adhered to.

Since air is a good insulator, overhead cables which are out of reach of people or animals do not need further insulation. The cables we have just described are sheathed, but Overhead Line Cable is also made consisting of bare copper or aluminium conductors of a size to have enough mechanical strength to support themselves. These cables can span long distances and are used by Electricity Boards for their distribution systems outside towns. They are used much more in public supply systems than in building services.

Whatever method of running overhead cables is used, at the end of the overhead section the cables must be connected to the cables within the building in a terminal block and in such a way that there is no mechanical strain on them.

Cable entries

The entry of cables from the outside to the inside of a building sometimes causes difficulty. There must obviously be a hole in the wall which has to be tight round the cable and which has to be sealed to prevent dirt, vermin and moisture getting in. Whether the cable is an armoured type laid directly in the ground or whether it is drawn into a duct the most practicable way of making the entry into the building is by means of an earthenware duct built through the wall below ground level. When the cable has to bend up to rise on the inside face of the external wall a duct bend can be built into the wall.

After the cable has been pulled through the earthenware duct a seal is made round it within the duct with a bituminous mastic compound. Normally this is inserted from the inside of the building. The essential requirement is to make the seal watertight; it will be readily understood that a seal which prevents water coming through will also stop dirt and small animals. In difficult cases one can make a metal plate to overlap the earthenware duct with a hole in it of a diameter to be a push fit on the cable. The duct is filled with mastic, the metal plate is pushed over the cable to cover the end of the duct and is screwed back to the wall, and the edges of the plate are then pointed with mastic. This construction gives an effective water seal.

Temporary installations

Temporary installations must be just as safe as permanent ones. There is therefore no reason for departing from any of the principles of design and installation which are used for permanent systems. The methods of cable sizing and schemes of distribution which are described in the following chapters apply to temporary installations as well as to permanent ones. The methods of installing cables which we have discussed in this chapter are all designed to give adequate safety and can be used on any temporary installation.

There is however one relaxation that can be accepted on temporary installations and that is that overhead cables need not be supported at such close intervals as permanent overhead cables. It is fairly common practice on building sites to run PVC insulated PVC sheathed cables overhead between site huts without intermediate support.

The almost universal way of wiring inside temporary huts is PVC sheathed cable exposed on the surface. Wiring from building site huts to local temporary power or lighting points on the site is also carried out in PVC sheathed cable. Where it cannot go overhead it should be properly clipped to suitable walls. Particular care must be taken to see that it is not in the way of cranes and lorries moving across the site.

This book deals with design rather than installation, and methods of wiring have been described from this point of view. More information on practical aspects of installation can be obtained from books intended for installation rather than design, such as *Modern Wiring Practice* 8th Edition by Steward & Watkins.

Standards relevant to this chapter are:

BS 31	Steel conduit and fittings
BS 731	Flexible steel conduit
BS 951	Earthing clamps
BS 2706	Non ferrous conduit and fittings
BS 4568	Steel conduit and fittings with metric threads
BS 4607	Non metallic conduit and fittings
BS 4678	Cable trunking

I.E.E. Wiring Regulations particularly applicable to this chapter are:

Regulation	521—9
Regulation	521—10
Regulation	521—11
Regulation	521—12
Regulation	521—13
Regulation	521—14
Section	523
Section	525
Section	526
Section	527
Section	528
Section	529
Regulation	553—10
Regulation	553—11
Regulation	553—12
Appendix	11
Appendix	12

4 Cable Rating

An important part of any electrical design is the determination of the size of cables. The size of cable to be used in a given circuit is governed by the current which the circuit has to carry, so the design problem is to decide the size of cable needed to carry a known current. Two separate factors have to be taken into account in assessing this, and the size of cable chosen will depend on which factor yields the larger value in each particular case.

A conductor carrying a current is bound to have some losses due to its own resistance. These losses appear as heat and will raise the temperature of the insulation. The current the cable can carry is limited by the temperature to which it is safe to raise the insulation. Now the temperature reached under continuous steady state conditions is that at which the heat generated in the conductor is equal to the heat lost from the outside of the insulation. Heat loss from the surface is by radiation and conduction and depends on the closeness of other cables and on how much covering or shielding there is between the cable and the open atmosphere. Thus the heat loss and, therefore, the equilibrium temperature reached depends on how the cable is installed; that is to say, whether it is in trunking, or conduit, on an exposed surface, how close to other cables, and so on. To avoid tedious calculations, tables have been prepared and published (chiefly in an appendix to the I.E.E. Regulations for the Electrical Equipment of Buildings) which list the maximum allowable current for each type and size of cable.

The tables give a current rating for each type and size of cable for a particular method of installation and at a particular ambient temperature. For these basic conditions a cable must be chosen the rated current of which is at least equal to the working current. For other methods of installation and ambient temperatures the tables give various correction factors. The actual working current has to be divided by these to give a nominal current and a cable then selected such that its rated current is at least equal to this nominal current.

Particular care has to be taken where cable is run in a thermally insulated space. With increasing attention to thermal insulation of walls this is likely to become a more frequently occurring situation, and the I.E.E. Regulations now require a cable to be de-rated when it is used in such a situation.

As explained in Chapter 9, every cable must be protected against overload. The working current must be such that if it is exceeded the resulting rise in temperature will not become dangerous before the protective device cuts off the current.

When a short circuit occurs the cable is carrying the fault current during the time it takes the protective device, whether a fuse or a circuit breaker, to operate and disconnect the circuit. Because this time is very short the cable is heated adiabatically and the temperature rise depends on the fault current and the specific heat capacity of the cable. The fault current depends on the earth fault loop impedance, which is explained in Chapter 9. This impedance is the sum of the impedance of the circuit cable and the impedance of the protective conductor. The I.E.E. Regulations give methods for checking that both these impedances are low enough for the protective device to operate before a dangerous temperature is reached.

In most cases the protective device will have a breaking capacity greater than the prospective short circuit current, and this allows one to assume that the current will be disconnected sufficiently quickly to prevent overheating during a short circuit. The cable size selected from the rating tables for the working current is then adequate.

In other cases a further check must be made by means of the formula

$$t = \frac{k^2 S^2}{I^2}$$

where t = time in seconds in which protective device opens the circuit at a current of IA

k = a constant, given in the Regulations for different cables

S = minimum cross sectional area of conductor in the cable, mm^2

I = short circuit current, A

If necessary the cable size must be increased above that provisionally selected from the tables in order to satisfy this condition.

Alternatively, the cable size can be retained and a fuse or circuit breaker with a faster operating time used.

The protection must also operate if the overload is not a short circuit but a comparatively small multiple of the working current. HRC fuses and circuit breakers can take up to 4 hours to operate at a current 1.5 times their rated current. The cable temperature will rise during this time and the working current must allow a safety margin to take account of this. The rating tables in the I.E.E. Regulations include the necessary margin for HRC fuses and circuit breakers.

However, rewirable fuses take longer to operate and a larger margin is therefore necessary. The rating tables therefore include a factor by which cables must be de-rated if rewirable fuses are going to be used to protect the cables.

The resistance of the conductor also results in a drop of voltage along its length. Because of this drop, the voltage at the receiving end is less than that at the sending end. Since all electrical equipment used in a building is designed to work on the nominal voltage of the supply in the building, it is necessary to limit the amount by which the voltage drops between the point of entry into the building and the outlet serving an appliance. In other words, the voltage drop in the wiring must be kept reasonably low. The I.E.E. Regulations require that the voltage drop in the wiring should not exceed 2.5% of the nominal voltage. For a 240 V supply, this is 6 V.

The drop in volts is obtained by Ohm's law as the product of the actual current flowing and the total resistance of the actual length of cable. One therefore wants to know the resistance per unit length of cable. The cable rating tables already mentioned give this, but for convenience in use instead of giving it as ohms per metre they quote it as voltage drop per amp per metre length of cable. This makes it a very simple and quick matter to calculate the actual drop over the actual length for the actual current. If this is more than the acceptable drop the next larger sizes of cable must be chosen and the calculation repeated. The author finds it quicker in practice to take the maximum allowable voltage drop, divide it by the length of cable and divide the quotient by the current to be carried. This gives the maximum permissible volts drop per metre per amp, and it is only necessary to look in the rating tables for the smallest size of cable that has a volt drop less than the figure calculated.

An example may make the procedure clearer. It was required to determine the size of cable necessary to be run along the side of a driveway to serve eight road lighting columns. Each column was to carry a lantern having a 35 watt SOX lamp taking 0.6 amps. The length of cable was scaled from the drawing of the scheme as 300 metres. The cable would come straight from a fuse switch at the main intake and the full drop of 6 volts could therefore be allowed in it.

Total current is 8 x 0.6 = 4.8 amps

For PVC SWA cable in a trench correction factors are:

for ambient temperature 25°C   1.06

for single cable   1.0

$\therefore$ nominal current rating of cable $= \dfrac{4.8}{1.06 \times 1.0} = 4.5$

The fuse switch at the intake will have an HRC fuse with a breaking capacity of 16 500 A. At the point in the supply undertaking's network from which the supply is taken the prospective short circuit current is not more than 10 000 A.

The breaking capacity of the protective device is therefore more than the prospective short circuit current and no further calculation is necessary.

Allowable voltage drop is 6000 mV over 300 metres

$$\therefore \text{ allowable drop is } \frac{6000}{300 \times 4.8} \frac{mV}{mA}$$

$$= 4.2 \frac{mV}{mA}$$

∴ cable must have a current rating of 4.5 A or more and volt drop 4.2 mV/mA or less

From the cable rating tables the smallest size cable which meets both requirements is 16 mm². In fact the voltage drop and not the current rating is the determining factor.

It can be seen that whereas the current carrying capacity, being limited by the temperature of the insulation, is a fixed property of the cable, the voltage drop depends on the length of the circuit. The former, therefore, sets an absolute maximum to the current which a given cable can carry, while the latter may, depending on the length of the circuit, restrict the permissible current to less than this maximum. Both factors have to be looked at by the designer.

In most cases, there will be a number of sub-mains from the electrical intake of the building to distribution fuse boards, and from each of these there will be a number of final sub-circuits. The allowable voltage drop is the sum of the drops in the sub-mains and in the final sub-circuits, and there is no restriction on how it is shared between the two. The position of each distribution board will affect the lengths both of sub-mains and of final sub-circuits and thus of the voltage drop in each of them. There is no single correct way in which these parameters must be combined, and the design can only be done by a process of trial and error tempered by the designer's own experience and judgement. There is plenty of scope for a designer to exercise his personal initiative and intuition in positioning distribution boards and selecting cable sizes to arrive at an economical design.

Mention should also be made of the protective or earth cable. The function of this is described in Chapter 9. Under normal conditions it carries no current and it conducts electricity only when a short circuit occurs and then only for the short time before the protective device operates. The I.E.E. Regulations give two alternative ways of determining its size. The first is by the use of the same formula as has been quoted above for checking the rating of the line cable.

Alternatively, the Regulations give a table which relates the size of the protective conductor to the size of the line conductor. The effect is that for circuits up to 16 mm² the protective conductor must be the same size as the line or phase conductor, for 25 mm² and 35 mm² phase conductors the protective conductor must be $16\ mm^2$ and $25\ mm^2$ respectively, and for phase conductors over $35\ mm^2$ the cross section of the protective conductor must be at least half the cross section of the phase conductor.

I.E.E. Wiring Regulations particularly applicable to this chapter are:

Section   522
Section   523
Regulation   543—1
Regulation   543—2
Regulation   543—3
Appendix   8
Appendix   9

5 Circuits

The final outlets of the electrical system in a building are lighting points, socket outlets and fixed equipment. The wiring to each of these comes from a fuse in a distribution board, but one fuse can serve several outlets. The wiring from one fuse is known as the final sub-circuit, and all the outlets fed from the same fuse are on the same sub-circuit. The fuse must be large enough to carry the largest current ever taken at any one instant by the whole of the equipment on that sub-circuit. Since the fuse protects the cables, no cable forming part of the circuit may have a current carrying capacity less than that of the fuse. The size of both fuse and cable is, therefore, governed by the number and type of outlets on the circuit.

It is unusual to have a fuse of more than 30 A in a final distribution board, and the cables normally used for final sub-circuits are 1.5 mm^2, 2.5 mm^2 and 6.0 mm^2, according to the nature of the circuit. Lighting is almost invariably carried out in 1.5 mm^2 cable and power circuits to socket outlets in 2.5 mm^2. 6.0 mm^2 cable is used for circuits to cookers and other large current using equipment, such as machine tools in workshops. These sizes are so usual that it is better for the designer to restrict the number of outlets on each final sub-circuit to keep within the capacity of these cables than to specify larger cables. If he does choose the latter course there is a real danger that the site electrician will install the cables he is used to, instead of complying with the designer's specification. These considerations will not, of course, apply in a factory in which individual machines can take very heavy currents.

It is worth noting here that a single phase 1 hp motor takes a running current of about 6 A and a starting current of something under 30 A. These currents would also be taken by each phase of a three phase 3 hp machine. Therefore, quite large machine tools impose no bigger a load than a cooker, and can be served from a fuseway on a standard distribution board. The distribution in a workshop or medium sized factory can follow the same principles as that in a domestic or commercial building. Indeed, in a technical college, the electrical load in the metal workshops can well be lower and impose fewer problems than that in the domestic science rooms with their many cookers.

For factories with heavier machinery, fuseboards with 60 A or 100 A fuses can be used, with correspondingly larger cables. For factories with very large loads, the normal type of distribution board ceases to be practicable and other means of arranging the connections to the machinery are adopted.

We have said that the fuse must be rated for the largest current taken at any one instant by all the equipment on the circuit. This is not necessarily the sum of the maximum currents taken by all the equipment on the circuit, since it may not happen that all the equipment is on at the same time. One can apply a diversity factor to the total installed load to arrive at the maximum simultaneous load. To do this, one needs an accurate knowledge of how the premises are going to be used, which one can get by a combination of factual knowledge and intuition. An appendix in the I.E.E. Regulations gives a scheme for calculating the diversity factor, and circuits designed in accordance with this will comply with the regulations; nevertheless the author of this book thinks that a capable designer will not rely on such a rigid guide to the total exclusion of his own judgement. A general knowledge of life and of how different buildings are used may be of more help than theoretical principles.

It will be easier to understand the ideas underlying the use of circuits intended for fused plugs if we first consider the limitations of other circuits.

The fuse in a final sub-circuit may not have a rating greater than that of the cables in the circuit. If it did, it would not protect the cables against overloads falling between the capacity of the cables and the normal current of the fuse. However, neither the designer of the building wiring nor the installer has any control over the sizes of flexible cables attached to portable appliances which will be plugged in at socket outlets. When electricity was first introduced for domestic use, the system adopted for house wiring in the United Kingdom was that 2 A socket outlets were provided to serve radios and portable lamps which would have small flexibles, 5 A outlets were provided for larger equipment and 15 A sockets for the heaviest domestic appliances such as 3 kW fires, which would be supplied with substantial flexible cables. If carefully designed and properly used, such a system would give reasonable protection, not only to the permanent wiring of the building, but also to the flexibles and portable appliances plugged in at the socket outlets. Unfortunately, the multiplicity of plugs and sockets made life difficult for the householder and tempted him to use multi-way adaptors, which totally defeated the object of having different sized outlets. Also, the use of electricity has greatly increased since its first introduction to domestic and commercial premises, so that it has become necessary to have a large number of socket outlets in each dwelling and in every office. With the original system of wiring, it would have been necessary to have many more fuses, the fuseboards would have become larger, more numerous, or both, and the cost of the installation would have increased rapidly.

It was to overcome these difficulties that 13 A socket outlets with fused plugs were introduced a few years after the Second World War. The socket outlets are made to BS. 1363 and the fuses that go in the plugs to BS. 1362. The fuse in the plug protects the flexible cable and the appliance connected to it. Any overload in the appliance or any damage to the flexible cable will blow the fuse in the plug; provided this fuse is correctly rated to protect the appliance, it does not matter if the fuse in the permanent wiring has a higher rating. This latter fuse now has to protect only the permanent house wiring up to the socket outlet. The permanent wiring can, therefore, be designed without consideration for protection of appliances, which have been given their own protection. There is no longer any need to have different outlets for different classes of appliances, and it is possible to standardize on one type of socket and one type of plug.

This system depends on matching the fuse in the plug to the appliance. Fuses for these plugs are made to BS. 1362 in ratings of 3 A, 5 A, 10 A and 13 A. It is unfortunate that many people do not realise this and that many retailers sell every 13 A plug with a 13 A fuse in it. The result is that many light current appliances, such as radio and TV sets, are not properly protected.

Fused plugs bring a further advantage. Whereas a 15 A socket had to be assumed to be feeding something taking 15 A, a 13 A socket with a fused plug may well be feeding equipment taking 2 A or less. Therefore, where there are several 13 A outlets, it becomes permissible to make use of a diversity factor in deciding on the circuit loading.

This is taken into account in the standard circuit arrangements given in the I.E.E. Regulations, which should be used wherever possible.

A ring circuit with socket outlets for 13 A fused plugs wired in 2.5 mm^2 PVC cable and protected by a fuse or circuit breaker rated at 30 A can serve any number of outlets but the floor area covered must not be more than 100 m^2. A radial circuit for this type of outlet can serve a floor area of 50 m^2 if it is wired in 4 mm^2 cable and protected by a 30 A HRC fuse or circuit breaker. If it is wired in 2.5 mm^2 cable and protected by any type of fuse or circuit breaker rated at 20 A it is restricted to a floor area of 20 m^2. In either case there can be any number of outlets within this area.

Any number of fused spurs may be taken from any of these three circuits, but the number of unfused spurs is limited to the number of socket outlets on the circuit.

The cable sizes quoted are for copper conductors with PVC insulation. The regulations give different sizes for MICC cables and for PVC insulated aluminium cables.

15 A and 5 A socket outlets must be assumed to supply appliances taking 15 A and 5 A respectively, and no diversity may be applied to circuits containing such outlets. The circuit may however contain any number of such outlets provided the rating of the protective device is equal to the sum of the ratings of

the outlets on the circuit. Thus a circuit with a 15 A fuse may feed one 15 A socket or three 5 A sockets. The framers of these regulations had in mind chiefly domestic installations, and the arrangement may be departed from in non-domestic installations if the exact usage is known. To illustrate this we can take as an example a hospital ward in which a socket outlet is needed next to each bed for portable cardiographs or X-ray equipment. It is also desired to ensure that only certain equipment can be connected to these sockets. This can be done by installing 15 A outlets and providing 15 A plugs only for the equipment which is to use these special outlets. It may be known that although there is to be one of these outlets next to each bed only one patient in a ward will be treated at one time. In such a situation it is clear that twenty socket outlets could be put on one circuit and yet that circuit would never carry as much as 15 A, and an exception from the standard circuit could therefore be made.

A cooker, whether served through a control switch or through a cooker control unit, should be on a circuit of its own. The cooker control unit may incorporate a socket outlet, which will be on the same circuit. The circuit rating should be that of the cooker. Two cookers in the same room may be on a single circuit provided its rating does not exceed 50 A.

We have referred to a ring circuit and we must now consider what this is. As its name implies, a ring circuit is one which forms a closed ring; it starts at one of the ways of a distribution board, runs to a number of outlets one after another, and returns to the distribution board it started from. This is illustrated in Fig. 53. The advantage of this arrangement is that current can flow from the fuseway to

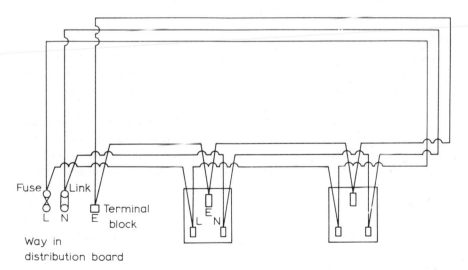

Fig. 53   Ring circuit

the outlets along both halves of the ring, so that at any one point the cable carries only part of the total current being taken by the whole circuit. It is this feature which makes it possible for the fuse rating to be greater than the cable current rating. The fuse carries the sum of the currents in the two halves of the ring and will blow when the current in one part of the ring is about half the fusing current of the fuse.

A circuit which runs only from the fuseway to the outlets it serves without returning to the fuse, is called a radial circuit, to distinguish it from the ring circuit which we have just described. Every circuit is necessarily either radial or ring.

In explaining the rules for the number of outlets on a ring circuit, we have spoken of spurs, and we should pause to explain what is meant by this. Ideally the outlets on a ring are placed so that the cable can run from the first to the second and from the second to the third without doubling back on itself. In some places, this may not be possible and the cable must return by the same route as it came, so that the ring is closed in on itself. If one outlet is a long way from the others, this doubling back may be expensive in cable and it may be cheaper to serve the odd outlet by a radial branch or spur from the ring. The reasoning which applies to the choice of cable and fuse size for the ring does not apply to the cable in the spur. This length of cable must be protected by a fuse with a rating not greater than that of the cable and as this will be less than the rating of the fuse in the ring circuit, there must be a fuse at the point at which the radial spur leaves the ring. A fused spur unit is a convenient device for providing this fuse. It will be seen from the rules given above that a considerable number of spurs may be taken from one ring, but in practice this is very seldom done.

Fused spur units are also used for connected fixed appliances to ring circuits even when they are close to the line of the ring. The fuse in the spur unit performs the same function as the fuse in the plug of a portable appliance and protects both the appliance and the short length of cable between the outlet and the appliance.

Although the ring circuit was developed for domestic premises, it is equally useful for commercial premises and is frequently used for the power wiring of offices and shops. In housing, it is standard practice to put all the socket outlets on one floor of a house on one circuit, but a little more thought is clearly needed in the layout of the circuits in commercial premises. The number of outlets on a ring is ultimately limited by the rating of the fuse. For premises other than factories, it is almost universal to run ring circuits in 2.5 mm^2 PVC cable and to fuse them at 30 A. The designer must assess the maximum current likely to be taken at any one time and plan such a number of separate circuits that none of them will be required to supply more than 30 A at a time. When doing this, he should remember that the use of electricity has increased enormously in the past few decades and is likely to go on increasing. It is, therefore, possible that within the lifetime of an installation more appliances will be plugged in simultaneously than is usual today, and also that individual appliances may be heavier users of current than the appliances in common use today. Some allowance must be made for a future increase of use, and in the absence of any other way of doing it, the author feels that it is prudent to restrict the present maximum current to say 15 to 20 A per ring circuit.

In a school or college workshop, it is often desirable for an emergency stop button to switch off all machines. The usual arrangement is that prominent stop buttons are fixed at two or three easily accessible places in the workshop so that in the event of any pupil having an accident the master in charge can stop the machine quickly wherever he himself happens to be at the time. All machines have to be controlled together so that if an emergency button is pushed they all stop. A further requirement for safety is that the circuit must be such that the machines will not start again until the emergency stop has been reset. There are basically two ways in which an emergency stop circuit to meet these requirements can be carried out.

If the machine tools are large and each takes a large current, it will be better to feed each one on a separate radial circuit from its own fuseway. There will then have to be a distribution board with a number of fuseways serving a number of machines. The incoming supply to this board can conveniently be taken through a contactor which is normally open but is held shut when the operating coil is energized. The circuit of the operating coil is taken round the workshop and goes through as many emergency stop buttons as are needed. The result is that when any one of these buttons is struck, the operating coil is de-energized, the contactor opens, and the entire fuseboard is de-energized. Everything fed from that board then stops.

The cables of the operating coil circuit must themselves be protected by a fuse. If the circuit is taken from the incoming supply to the contactor, if it is to rely on the fuse protecting the main incoming circuit, it must be of the same size as the cables of the incoming supply. Normally, these cables will be far heavier than is necessary for an operating coil circuit, but if smaller cable is used for the latter, the fuse protecting the incoming supply will be too large to protect the smaller cable. In that case, a fuse unit must be inserted where the operating coil circuit branches from the main supply.

A large workshop may have so many machines that it requires two or three distribution boards to supply them all. The emergency circuit must then shut off the supply to all of these boards. It would be possible to take the sub-mains to several distribution boards through a multipole contactor with one pole for each live and each neutral, but such contactors are not readily available, and it is better for each distribution board to be fed through its own contactor. The emergency stop circuit can contain a relay with one pole for each contactor, and the operating coil circuit of each contactor is then broken by the relay when the relay itself is de-energized on the interruption of the emergency circuit. This is a simple arrangement, but needs an extra circuit for the relay, and this circuit cannot come from any of the distribution boards which the relay controls. One way of avoiding the extra circuit and the relay would be to wire the operating coils of the contactors in parallel so that they were all part of a single circuit. This would have the disadvantage that at least one of the contactors must have its coil and main contacts fed from two different sources, so that it would be possible for the coil to be alive when the main feed to the contactor had been disconnected. Such an arrangement can cause damage to unwary maintenance electricians and is not recommended. It is safer to pay for the relay.

Similar circuits can be adopted whenever it is necessary to control a large number of points together or from a remote place. The external lights of a hotel or public building may, for example, be sufficiently extensive to require several ways of a six or eight distribution board. If the distribution board is controlled by a contactor, all the lights can be switched together on one switch, which is in the operating coil circuit of the contactor.

The second method of providing an emergency stop circuit is appropriate for smaller workshops in which it may be cheaper and quite satisfactory to serve all machines from a single ring circuit. The outlets on the ring main take the form of fused isolators, and each machine is connected locally to its own fused isolator. The emergency circuit still works a contactor, but in this case, the contactor is on the load side of the fuseboard. The supply is taken from the fuse to the contactor, and the ring starts from and comes back to the contactor. The operation of the emergency stop cuts out everything fed from this one fuseway but leaves in operation circuits from all the other ways of the distribution board. For a small workshop, this saves the expense of a separate distribution board for the machines only. The coil circuit must either be of the same size as the cables of the ring main or have its own fuse.

We have so far discussed mainly power circuits and must now say something about lighting circuits. It is usually necessary to have several lights on one circuit with each light controlled by its own switch. Fig. 54 shows the wiring arrangement used to achieve this. It also shows circuits for two way, and for two way and intermediate switching.

Examination of these diagrams will reveal that the flexible cord to the lampholder is protected by the fuse in the whole lighting circuit. This fuse must not, therefore, be of a higher rating than the smallest flexible cord used on that circuit. In domestic premises flexible cords will almost inevitably be 0.75 mm^2 rated at 6 A and so domestic lighting circuits should not be fused at more than 5 A. Now a 150 W tungsten bulb takes 0.63 A and consequently a 5 A circuit can have eight of these on simultaneously. In an average house, the 5 A limit will not be exceeded if there is one lighting circuit for upstairs and one for downstairs, but in a large house it may be necessary to have more lighting circuits than this.

In commercial and industrial premises far too many circuits would be needed

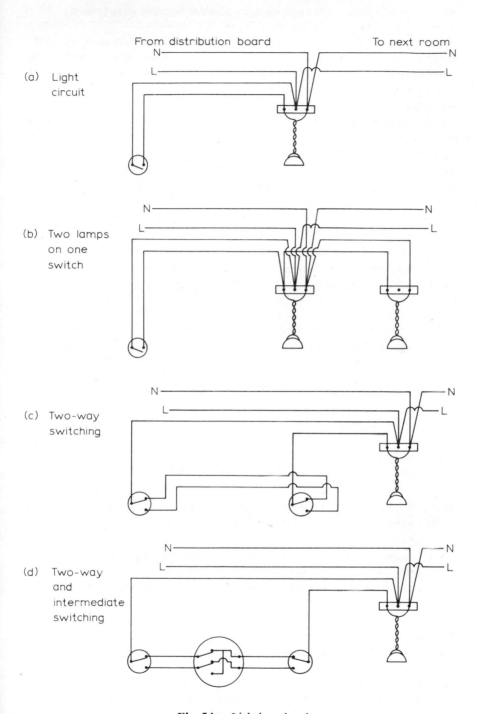

(a) Light circuit

(b) Two lamps on one switch

(c) Two-way switching

(d) Two-way and intermediate switching

From distribution board

To next room

Fig. 54 Lighting circuits

if they were all restricted to 5 A. There is no objection to lighting circuits being rated at 10 or even 15 A, but the designer and installer must make sure that all pendant cords to lights are of the same rating as the rest of the circuit.

It has been explained in Chapter 1 that in the wiring of buildings cable joints are not made by soldering but by mechanical connectors. It is a help to maintenance if loose connector blocks can be avoided and cables joined together only at the various outlets. Furthermore, each connector adds a small joint resistance and it is advantageous to keep the number of these down to a minimum. Mechanical connections cannot be avoided at the outlets, but the number of joints can be reduced if no connections are made except at the outlets. This method of joining cables only at outlets is known as the 'looping in' system, and the diagrams in Figs. 53 and 54 show how it is achieved. One piece of cable runs from the first outlet (socket outlet or ceiling rose as the case may be) to the second outlet; if the wiring is properly planned and carried out there need be no join in this length. A second piece of cable runs from the second outlet to the third. Both these cables connect to the same terminal at the second outlet and

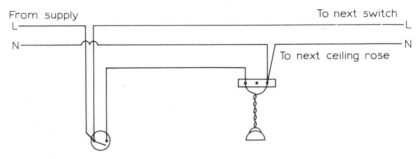

Fig. 55   Looping at switch

thus the circuit becomes continuous. No joints are needed except at the outlets.

The looping in system is the reason for having the third terminal on a ceiling rose, which we described in Chapter 1. It can be seen from Fig. 54 that the third terminal is needed to join the incoming and outgoing live wires. It remains alive when the light attached to the rose is switched off, and can give an unpleasant surprise to the home handyman replacing a light fitting. All good ceiling roses have this terminal shrouded to prevent accidental contact, but nevertheless, some specifications insist that only two terminal roses be used. The looping could be done at the switch, as shown in Fig. 55, but it is difficult to visualize a building in which this scheme would not require very much more cable than that of Fig. 54. The only practicable way of using a two terminal rose is to make the third joint with a connector block fitted loosely in the conduit behind the rose. This is clumsy, and the author's preference is for a three terminal rose with the third terminal properly shrouded.

We have described the way in which outlets are conventionally grouped and arranged in final sub-circuits. Each of these sub-circuits is fed from a fuse on a distribution or fuseboard and the next step in describing a complete electrical system is to show where the distribution board gets its supply from. This we shall do in the next chapter.

I.E.E. Wiring Regulations particularly applicable to this chapter are:

Section   314
Section   463
Appendix   4
Appendix   5

6   Distribution

Electricity is supplied to a building by a supply authority; in the United Kingdom this is an Area Electricity Board, while in other countries it may be an Electricity Supply Company or public body. The supply is provided by a cable brought from outside into a suitable point in the building, which is referred to as the main intake, and from this electricity has to be distributed to all outlets which use it. The incoming cable may be a 120 or 150 mm^2 paper insulated cable and the current flowing along it must be divided between a number of smaller cables to be taken to the various final destinations throughout the building. This division is the function of the distributing system.

In Chapter 5, we described the final sub-circuits which serve the final outlets. Each final outlet takes a comparatively small current, and it would be impracticable to serve it with a large cable. As we have seen, the final sub-circuits are most commonly wired in 1.5 mm^2, 2.5 mm^2 and 6 mm^2 cables. The cable size in turn limits the number of outlets on each circuit, and in a building of any size a large number of circuits is needed. It would be very expensive to run all the final sub-circuits from the main intake point. Also, voltage drop in cables of this size over a long distance would be excessive. It is more economic and more practical to divide the supply first over a few large cables and then into the final small cables in a second step. The normal method is to distribute current from the main intake to a number of distribution or fuseboards, each of which splits it further among a number of final sub-circuits. A typical scheme is shown diagrammatically in Fig. 56.

The cable from the main intake to a distribution board is known as a sub-main, and it must be rated to carry the maximum simultaneous current taken by all the final sub-circuits on that board. Once this current is known, the size of the cable can be determined by current carrying capacity and voltage drop, as explained in Chapter 4.

The fuses in the distribution board protect the final sub-circuits, but the

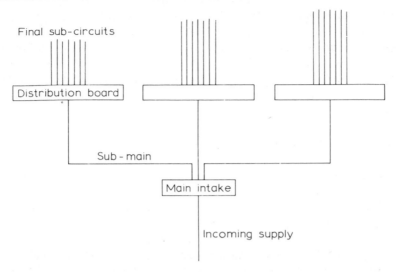

Fig. 56   Distribution

sub-main cable also needs protection against short circuits and overloads, and there must be a fuse or other protective device at the main intake. We can note the principle that every cable must be protected at its feeding end. A convenient device for protecting a sub-main cable is a switch fuse. A switch fuse is illustrated in Fig. 57, and it can be seen that it consists simply of an isolating switch and a fuse carrier housed together in a substantial casing. The one illustrated is a three phase switch fuse and, therefore, has three poles on the

Fig. 57   Switch fuse (*Courtesy of* Bill Switchgear Ltd.)

switch and three fuses. The fuse protects the cable leaving the switch fuse and the switch is useful for isolating the sub-main from the rest of the electrical system when this is required for maintenance or alteration work. The switch is designed to make and break the rated current of the switch fuse, and units are made in a series of standard ratings. A typical model, for example, is manufactured in increasing sizes rated at 30, 60, 100, 160, 200, 400, 600 and 800 A.

A switch fuse also includes terminals which enable the earth wires on the incoming and outgoing sides to be connected together. Under no circumstances must there be a break in this circuit as it would destroy the safety of the system. The neutral wire on the other hand can be taken through the switch fuse in one of two ways.

The more usual way is for the switch fuse to include terminals for connecting the incoming and outgoing neutrals in the same way as the earth wires. The alternative is for it to have a switch blade in the neutral line as well as in the live line. In this case there is a solid link instead of a fuse in the neutral line.

A fuse switch, illustrated in Fig. 58, is similar to a switch fuse, but in this case the fuse carriers are mounted on the moving blades of the switch.

The whole of the current going into the sub-main passes through the switch fuse which carries no current for any other part of the system The total incoming current must be divided to go to several switch fuses, and the simplest device for distributing current from one incoming cable to a number of outgoing ones is a busbar chamber. This consists of a number of copper bars held on insulating spacers inside a steel case. It is shown in Fig. 59. Cables can be connected to the bars anywhere by means of cable clamps which are usually

Fig. 58   Fuse switch (*Courtesy of* Bill Switchgear Ltd.)

Fig. 59   Busbars (*Courtesy of* Bill Switchgear Ltd.)

screwed to the bars. The incoming cable can be connected to the bars at one end or at some convenient point along them, and the outgoing cables are connected at suitable intervals along them. The switch fuses are usually mounted immediately above and below the busbar chamber so that the connections, or tails, from the busbars to the fuses are kept as short as is reasonably possible. These tails are of the same size as the cables leaving the switch fuse but are protected only by the fuse on the main intake. It is not normally permissible to protect a cable with a fuse rated at more than maximum current carrying capacity of the cable, since this would leave the cable unprotected against any overload falling between its own capacity and the rating of the fuse. It may, however, be done for short tails between busbars and switch fuses. A short circuit or fault beyond the switch fuse will blow the switch fuse and thus stop current through the tail, and a short circuit on the tail will produce such a heavy overcurrent that the main intake fuse will blow. The only thing the tail is not protected against is a high resistance short circuit on itself which will produce a low overcurrent on itself but not beyond it. The probability of a high resistance fault on a short length of cable in the main switchroom is low enough to be neglected. Nevertheless, it should be neglected knowingly and the tails kept as short as possible.

Another matter which requires attention in the intake room is metering. The supply authority will want to meter the supply afforded and will want to install a meter at the intake position. If it is a small enough supply, the whole of it can be taken through the meter. The arrangement then is that the incoming cable goes first to a fuse which is supplied, fixed and sealed by the Electricity Board, and from there to the meter, which is also supplied and fixed by the Electricity Board. The fuse is the Electricity Board's service cut out and it is sealed so that only they have access to it. From the meter, the cable goes on to the busbar chamber. It is normal for the installer to provide the last piece of cable from the meter to the busbars but to leave the meter end of it loose for the Electricity Board to connect to the meter.

If the building takes a very large current, a different arrangement is used. It is not practicable to take a large current directly through a meter and large supplies are metered with the help of current transformers. The arrangement is similar to that described in the last paragraph, but the place previously occupied by a meter is taken by the primary coil of a current transformer. The output from the secondary coil of the transformer is taken to the meter, which can be made to give a direct reading of the current in the primary coil by suitable calibration. Some current transformers are made to slip over the busbars and use the bars as a primary coil. If this type is used the incoming cable must run from the service cut out to one end of the busbars and a space must be left on the bars between the incoming cable and the first outgoing cable. The method to be used should be agreed with the Electricity Board before installation commences even if this is one or two years before the building is to be finished and a supply of electricity will be needed. It is very embarrassing if, when the Electricity Board come to connect the supply, they find that there is not enough space for their equipment and it is much better to agree everything well in advance.

The size of busbars is determined by the current they are to carry which is normally the whole current of the building. The current carrying capacity is governed by temperature rise and has been tabulated in published data in the same way as the current carrying capacity of cables. The spacing between the bars is determined by the voltage at which they are to operate, since the air gap between adjacent bars and between bars and case is the only insulation provided. Busbars must also be capable of taking short circuit currents for the time it takes for fuses or circuit breakers to operate. In a short time, the bars will not overheat and the short circuit capacity is a measure of the mechanical forces which the bars will withstand. A heavy current gives rise to large electro-mechanical forces and the bar supports have to be capable of withstanding these. Busbars are obtained from specialist manufacturers and the electrical services engineer does not usually design his own. Further details on the methods of calculation would, therefore, be beyond the scope of this book.

A busbar chamber with a large number of switch fuses takes up a lot of wall space. It can also look untidy. Both these disadvantages can be overcome by the use of a cubicle switchboard. Such a board contains the busbars and the switch fuses all housed together in one large panel, a typical example being shown in Fig. 60. It works in exactly the same way as the busbar chamber with separate switch fuses, but is made in the manufacturer's factory instead of being assembled on site, and all the interconnecting wiring is inside the casing of the switchboard. This makes it possible to arrange the switch fuses in a more compact way and to get all the equipment into a much smaller space.

The meters can also be included within the composite switchboard. It is usual to have an incoming isolator or switch fuse which both protects the board against short circuits and makes it possible to isolate the board for maintenance.

Such a switchboard does, however, have disadvantages. Firstly, although it

Fig. 60 Cubicle switchboard (*Courtesy of* Bill Switchgear Ltd.)

requires less space than a site assembly of individual pieces of equipment, it is likely to cost more. Secondly, it is not easy to add further switch fuses to it once it has been made. A main intake consisting of separate pieces of equipment can be very easily extended; it is a simple matter to make an extra connection to the busbars and to take another pair of cables out through a short length of conduit to a new switch fuse fixed to any free wall space in the intake room. This is often necessary during the life of a building, and sometimes even before a new building is completed since building owners are apt to change their minds and want equipment installed which they had not thought of when building operations started. Unless blank spaces have been left in a cubicle switchboard it cannot be extended to take more switchfuses, and even if spaces have been left they often turn out to be not quite large enough for an unforeseen and unexpected extension.

In general each sub-main goes from the main intake to a distribution board. This consists of a case inside which is a frame holding a number of fuse carriers. Behind the frame, or sometimes alongside it or above it, is a busbar to which the incoming sub-main is connected. From the bar, there is a connection to one side of each fuseway provided. Each final sub-circuit is then connected by the installer to the outgoing terminal of one of the fuseways. The circuit is completed when a fuse carrier with a fuse is pushed into the fuseway. A second busbar is provided to which the incoming neutral and the neutrals of all outgoing circuits are connected. A typical distribution board is shown in Fig. 61. The only difference between a distribution board and a fuseboard is the name.

Standard distribution boards usually have either 4, 6, 8, 12, 16, 18 or 24 fuseways. Both single and three phase boards are available, the latter having

Fig. 61   Distribution board (*Courtesy of* Bill Switchgear Ltd.)

three fuseways for each outgoing circuit. It is not necessary to utilize all the available fuseways on a board, and in fact it is very desirable to leave several spare ways on each board for future extensions. These are often required before a building is even finished, and are almost certain to be wanted during the life of an installation which may last 40 to 60 years. A label should always be provided inside the cover of every distribution board stating which fuse serves which outlets.

The position of distribution boards within a building must obviously depend on the plan of the building. Apart from architectural considerations, it is a matter of balancing the lengths of sub-mains and the lengths of final sub-circuits to find the most economic way of keeping the total voltage drop between intake and final outlet to a minimum It is possibly better to keep sub-mains long and final sub-circuits short, but it is also desirable to keep the number of distribution boards down by having a reasonably high number of final sub-circuits on each board. To achieve this without excessively long final sub-circuits one must have the board fairly central for all the circuits it is serving.

In some cases, it is convenient to have a subsidiary control centre between the main intake and the distribution boards. When this is done a main cable runs from the main intake to the subsidiary control centre which is itself similar in construction to the main intake. The main cable is supplied through a switch fuse or fuse switch at the main intake end and leads to busbars at the subsidiary centre. The arrangement is shown schematically in Fig. 62. Such a scheme would be adopted only in a large building, but is very useful when several distribution boards have to be placed a long way from the main intake. It is more economical to keep the voltage drop down with one large main cable for the greater part of the distance to be covered than with several sub-mains running next to each other along the same route. It also requires less space for the cables. It is particularly useful when the premises being served consist of several different buildings. Normally, the Electricity Board will provide only one incoming service to one set of premises and if there are several buildings, the distribution to them must take place on the consumer's side of the meters. It is very convenient to run one main from the intake to a centre in each building and then distribute in each building from its own centre. Colleges and hostels are examples of consumers who may have several buildings on one site all forming part of the same premises.

The subsidiary control centre is also a good solution to the problem of extending existing buildings. A great deal of new building in fact consists of extensions to existing premises, sometimes by actual enlargement of an existing building or erection of a new building on spare ground within the site of the existing one. It is comparatively easy and does not need much room to add one switch fuse at the main intake. From this, it is again a comparatively simple matter to run one new main cable to the new building. Here it is possible to plan a subsidiary distribution centre with as many switch fuses and sub-mains to as

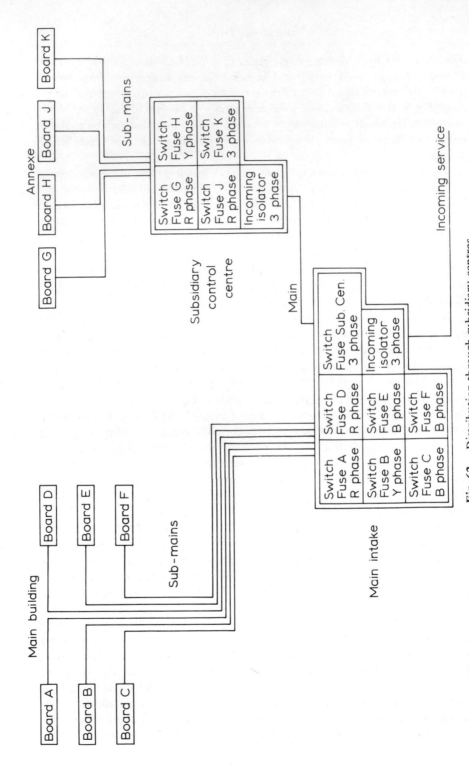

Fig. 62 Distribution through subsidiary centres

many distribution boards as are necessary. In this way, the amount of alterations in the existing building is kept to a minimum, and the new building is treated as an entity in itself.

When an extension is made it may be necessary for the Electricity Board to increase their service cable. This will depend on the existing load, the new load and the margin by which the capacity of the existing service cable exceeds the existing load. Whenever extensions are planned the capacity of the service cable must be checked with the supply authority, but replacing an incoming service with a larger one is not a difficult operation and does not add greatly to the work which has to be done in the existing part of the building.

In the United Kingdom, the Electricity Board's final distribution network to their customers is at 415 V three phase, but all domestic and nearly all commercial equipment requires a single phase 240 V supply. Motors above 3 h.p. are now very often three phase and they can be found in boiler rooms, kitchen ventilation plant, air conditioning plant rooms, lift motor rooms and similar places in buildings such as office blocks, schools and colleges, hospitals and blocks of flats. Any equipment of this nature will usually need a three phase supply, but distribution to lights and to ordinary socket outlets for power purposes must be at 240 V single phase. Whenever there is three phase equipment in a building the supply authority must obviously be asked to bring in a three phase supply. Where there is no three phase equipment in a building single phase supply would be enough, but there is another consideration which has to be taken into account. The supply authority wants its load to be evenly spread over the three phases of its network and this may need the co-operation of the consumers. For individual dwellings, whether flats, maisonettes or houses, the supply authority will bring in only a single phase cable, but it will arrange the supplies to adjacent or nearby dwellings so that the total supply is balanced over the three phases. Thus in a new development either each dwelling will be on a different phase from its immediate neighbour or small groups of dwellings will be on different phases from adjacent groups. For larger buildings, the Electricity Board will demand the co-operation of the consumer, or more properly of the engineer who plans the installation. The Board will insist that the consumer accepts a three phase supply even if he has no three phase equipment which he wishes to connect, and it will further insist that his demand is as nearly as possible balanced over the three phases. The designer must take some care about how he achieves this balance and he has to fulfil other requirements at the same time.

Most people are used to an electric supply at 240 V in their homes and expect the same in offices and public buildings. This is quite high enough to be dangerous, and it is largely for reasons of safety that the United States and some other countries have standardized domestic supplies at 110 V. It would be most undesirable to expose people to even accidental contact with 415 V and it is preferable to keep the three phases away from each other. In other words, the three phase supply comes in to the main intake and if necessary to subsidiary control centres but sub-mains and distribution boards are all single phase. An exception must, of course, be made for any sub-mains and distribution boards serving three phase machinery. Single phase outlets near each other should be on the same phase, and distribution boards close to each other should also be on the same phase.

The I.E.E. Regulations, which are almost invariably made to apply to installations in this country, stipulate that where fixed live parts between which there is more than 240 V are inside enclosures which although separate from each other are within reach of each other, a notice must be displayed giving warning of the maximum voltage between the live parts. Within reach is usually taken to mean 6 ft or 2 m It is clearly not pleasant to have a lot of notices in a building saying 'Danger — 415 volts' and one therefore plans the distribution within a building so that circuits on different phases are kept more than 2 m away from each other.

A very convenient way of doing this is to divide the building into three zones each of which is served by one phase. These zones must be of such sizes that each takes approximately the same load, in order that the total load is spread as nearly as possible equally over all three phases. The zones are not necessarily of the same area; for example a three storey school may have classrooms on two floors and laboratories on the third and the laboratories may account for more than a third of the total load. In such a case, a convenient division could be half the laboratories plus a third of the first floor on the red phase, the other half of the laboratories plus a further third of the first floor on the yellow phase, and the remaining third of the first floor plus the whole of the ground floor on the blue phase, but obviously the division must depend on the building and no general rules can be laid down. The designer must have enough personal judgement to settle each case on its merits.

It is helpful to label the three zones on a plan of the building. This will avoid confusion and draw attention to anomalies. It has the added and very important advantage of making the designer's intentions clear to the workmen on site and so reducing the risk of mistakes.

When a building is zoned it is worth checking all two way light switches. Suppose, for example, that ground and first floors are on different phases, and that the light on the stairs is wired on one of the first floor circuits. If it has two way switching the downstairs switch may be next to a corridor light switch which is on a ground floor circuit and, therefore, on a different phase. This should be avoided and is best prevented by careful labelling of circuits and phases on the drawings during design.

Even where three phase equipment has to be connected, the rules for separation of phases should still be applied to all single phase outlets near the three phase equipment. Operators and maintenance workers will be expecting a high voltage on the three phase equipment, but will not be expecting a high voltage to appear between a pair of single phase outlets near it.

Another point which has to be considered when the distribution through a building is being planned is the position of other services. If a fault develops on an electric cable its sheath or protective casing could become live. The various methods of giving protection against the consequences of such faults are described in Chapter 9, but the dangers arising from cable faults should also be guarded against when the cable routes are selected. If a live conductor breaks and touches the outer sheath or conduit the exposed parts become live. The protective device should immediately cut off the supply, as is explained in Chapter 9, but in the time it takes the protective device to operate, even if this is only a matter of milliseconds, there could be a spark from the damaged cable to adjacent metal. Should the adjacent metal be a gas pipe or some other service pipe there could conceivably be a serious accident.

For this reason good practice requires electrical wiring to be separated from other services. There should be at least 150 mm between them. If this is not possible the metalwork of the other services should be bonded to the earth continuity conductor of the electrical supply. This ensures that whatever happens the other service is at the same potential as the earth safety conductor so that no current or spark can pass between them.

The tariff which the Electricity Board will apply should be discussed with them when the installation is being planned. There are many tariffs available, and they differ from Board to Board. The consumer may find that he can obtain his electricity more cheaply by applying for a tariff which employs different rates for current used for power and for light. If this is so, the power and the light must be metered separately, and the designer must keep power and light circuits separate. Clearly he will want to know whether he must do so before he starts his work and he can only find out by making negotiations with the Electricity Board the very first part of his job.

If separate metering is decided on, it will be necessary to use split busbars at the intake. These are simply two lengths of busbars separated by a short distance from each other, but contained within the same casing. Externally, the appearance is the same as that of a single metered system, but internally the connections to the light and power systems are kept separate. The Electricity Board, of course, supplies two meters. Cubicle type main panels can be similarly arranged to keep the two services separate.

Each distribution board must be either for power only or for light only and must be fed by a sub-main from the appropriate set of busbars. In a large building with many outlets and, therefore, a considerable number of final sub-circuits, it is normally quite easy to have separate boards for light and power. For example, a technical college may need perhaps two lighting circuits and two power circuits in each laboratory, and a group of four laboratories may be so arranged that they can be conveniently served from one place. A pair of twelve way distribution boards in this place will provide the additional circuits needed for corridors and leave reasonable spare capacity. As two boards would be needed in any case and there are almost equal numbers of lighting and power circuits to be accommodated, it is quite convenient to have one board for power only and one board for lighting only. In small buildings, on the other hand, separation of light and power may make it necessary to use more distribution boards than would otherwise be required. Thus a sports hall may have four lighting circuits and only two socket outlets which can go on one power circuit. A six or eight way board would conveniently serve them all, leaving some spare capacity for future extensions. If the power is metered separately, a separate board will be needed for the single power circuit and it will probably have to be a three way board because this is the smallest size commercially available. A second sub-main will also be required and consequently an additional switch fuse at the intake. The result is a substantial increase in the cost of the installation, and the design engineer should keep this in mind when he is assessing the relative merits of different tariffs.

We have so far been talking in terms of distribution within a single large building. The same principles apply to distribution to dwellings but are modified by the fact that each occupier is a separate customer of the supply authority and must have his supply metered separately. In a suburban development of detached and semi-detached houses, the supply authority will bring a single phase service cable into each house. The number of circuits in a house is small enough to be accommodated on a single distribution board so that there is no need for sub-main distribution. The scheme thus becomes as shown diagramatically in Fig. 63. The incoming cable goes to a service cut out which is supplied, fixed and sealed by the Electricity Board. As a domestic supply does not normally exceed 100 A the whole can go through the meter and there is no

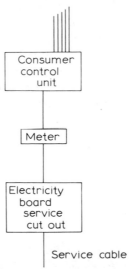

Fig. 63   Domestic service

occasion for the use of transformers. From the meter the cable goes straight to the distribution board. This must be supplied and fixed by the installer, who also provides the short tail to connect it to the meter. He leaves the tail loose at the meter for the Electricity Board to make the connection into the meter. In the case of a serious fault the service cut out will disconnect the consumer from the service cable and thus prevent the fault's affecting the rest of the Board's distribution network. It is obviously convenient to have all these pieces of equipment close together.

There are distribution boards made specially for domestic use, and these are known as Consumer Control Units, abbreviated to CCU. They are very similar to ordinary fuseboards but are single phase only and incorporate a double pole isolating switch on the incoming side. This makes it possible to cut off the supply from fuses before changing them. On a larger building, it is not so necessary to have an isolator on each board because the sub-main and board can be cut off together at the intake end. CCU's are available with 60 A or 100 A isolators and are made with up to 12 fuseways. A fairly typical arrangement of circuits for a semi-detached house is given in Fig. 64a and for a flat in a block with central heating in Fig. 64b. If a house requires a bigger supply than 100 A, it cannot be wired through a domestic CCU and it must be treated as a small commercial building with an intake put together from standard commercial fuse and distribution gear.

In a block of flats the arrangement within each flat is the same as that within a house, as just described, but some means must be found to take the supply

(a)	Circuits	Serving	Rating amps
	1	Upstairs lights	5
	2	Downstairs lights	5
	3	Garage and outside W.C. lights	5
	4	Upstairs ring main	30
	5	Downstairs ring main	30
	6	Immersion heater	15
	7	Cooker	45
	8	Spare	—

(b)	Circuits	Serving	Rating amps
	1	Lights	5
	2	Ring main	30
	3	Cooker	30
	4	Bathroom ventilation fan	10
	5	Clothes dryer	10
	6	Motorized valve on heating	10
	7	Spare	—
	8	Spare	—

Fig. 64   Domestic circuits

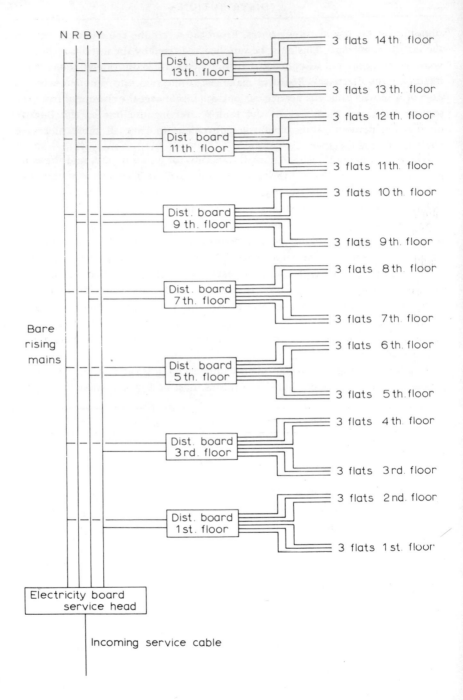

N R B Y

Dist. board
13th. floor

3 flats 14th. floor

3 flats 13 th. floor

3 flats 12 th. floor

Dist. board
11 th. floor

3 flats 11th. floor

3 flats 10 th. floor

Dist. board
9 th. floor

3 flats 9th. floor

3 flats 8 th. floor

Dist. board
7 th. floor

3 flats 7th. floor

3 flats 6th. floor

Dist. board
5 th. floor

3 flats 5 th. floor

3 flats 4 th. floor

Dist. board
3rd. floor

3 flats 3rd. floor

3 flats 2 nd. floor

Dist. board
1st. floor

3 flats 1st. floor

Bare
rising
mains

Electricity board
service head

Incoming service cable

Fig. 65 Distribution to flats

authority's cable through the block to each flat. In a small block, it may be practicable to bring into the block as many cables as there are flats and run them within the block to the individual flats. In this case, they will probably be PVC/SWA/PVC cables or perhaps paper insulated cables, but in a block of any size this would be a cumbersome solution, and the distribution on the board's side of the meter is similar to that which is used on the consumer's side of the meter in commercial buildings. The supply authority brings in one service cable to a service head which contains a fuse. From this, it takes a main cable feeding a series of distribution boards, and from each way on a distribution board it takes a sub-main into one flat. The flat is protected by the fuse on the distribution board and, therefore, no service cut out is needed in the flat itself.

Usually there is only a small number of flats on each floor and they can be served individually from a single distribution board. Successive floors in a block of flats are either identical or very similar, so that the various distribution boards are vertically above each other, and the main to them rises vertically. A convenient and very common way of carrying out the vertical distribution is by means of bare rising mains of the type illustrated in Fig. 26 (Chapter 2). A distribution scheme of this type is shown diagrammatically in Fig. 65, and a typical floor plan is shown in Fig. 66. It will be seen from Fig. 65 that the rising

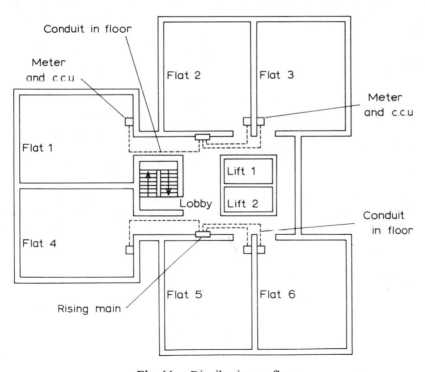

Fig. 66   Distribution to flats

main carries all three phases, but that each distribution board is connected to only one phase. In this way, the flats are allocated between the phases so as to make the total load as nearly balanced as possible. It has to be assumed that all flats impose the same load.

In this example, the Electricity Board service cable enters the building through an earthenware duct through the foundations, and terminates in a service head in a cupboard recessed in a wall at ground floor level. From this, the bare rising mains, enclosed in a sheet steel case, go up through the building. The architect has chosen to have the enclosing case semi-recessed in the corridor wall, but it could equally well be fully recessed, completely hidden inside a builder's work duct or completely exposed. On each alternate floor a six way distribution board is attached to the rising main bars. The system used enables the board to form part of the enclosing case. From each distribution board three 25 mm conduits rise to the floor slab above and three drop to the slab below, and within the slabs these conduits continue to a position just inside each flat. They terminate where the meter will be fixed and the CCU is fixed next to the meter. A three core 25 mm^2 PVC insulated cable is pulled through this conduit from the distribution board to each meter position.

It will be noticed from the first floor plan that there are actually two identical distribution systems within the building, each serving half the flats on each floor. This was done to avoid long and difficult horizontal runs at each level and also to keep the load on each rising main to below 400 A per phase. The whole of this distribution system is the property of the Electricity Board, who remain responsible for its maintenance. It may be put in during construction of the block by the Board's own staff or by contractors employed by the developer, in which case the contractors must work to the Board's specification. Which method is to be employed must be agreed beforehand by the developer and the Board; the plan of the installation must also be agreed between them when the building and its electrical system is being designed.

People do not stay at home all day and it is often difficult for meter readers to get in to read the meter. It is an advantage if the meter can be read from outside the dwelling, and on many new developments the supply is arranged to make this possible. One method is to have the meter in a purpose made cupboard in the outside wall of the dwelling. The CCU can be in an adjacent cupboard on the inside of the dwelling. Alternatively the CCU can be in any other convenient position inside the dwelling and instead of a short tail from meter to CCU there is a fairly long piece of cable. As this is likely to be at least a 25 mm^2 cable which cannot be bent too sharply the route it is to take should be planned beforehand to make sure the cable can be drawn in. In the case of a block of flats, instead of having the meter outside each flat, all the meters can be grouped together in some convenient central place. Fig. 67 is a modification of Fig. 66 with external meters. The six meters served by each distribution board are housed in a meter cupboard next to the distribution board, each meter being

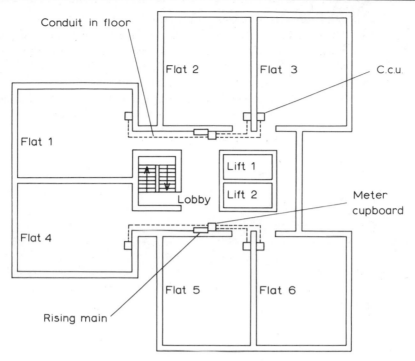

Fig. 67 Plan with external meter cupboards

connected by a short tail to the corresponding fuse in the distribution board. The conduit to each flat runs from the meter cupboard instead of from the distribution board, and goes straight to the CCU in the flat.

The second way of reading meters from outside is to use repeaters or slave indicators. A repeater, sometimes known as a slave indicator, is an indicating dial, identical to that in the meter itself, driven remotely from the meter. If this method were adopted in the example already discussed, the meters would be within the flats as shown in Fig. 66. Next to the distribution board would be a cupboard containing six repeaters, which would take the place of the meter cupboard shown in Fig. 67. 25 mm^2 cable would be taken through the conduit from the distribution board to each meter. An additional 1.5 mm^2 cable would run in the same conduit back from the meter into the distribution board and through the side of the board into the repeater cupboard. Here it would be connected to the repeater belonging to the flat in question, and the repeater would reproduce the reading on the meter.

Standards relevant to this chapter are:

BS 214   Distribution fuse boards for low and medium voltages
BS 1454   Consumer electricity control units

BS 4649   Miniature circuit breaker distribution boards

BS 5419   Specification for air break switches, air break disconnectors, air break switch disconnectors and fuse combination units for voltages up to and including 1000 V a.c. and 1200 V d c.

BS 5486   Factory built assemblies of switchgear and control gear

BS 6121   Mechanical cable glands for elastomer and plastics insulated cables

BS 6480   Impregnated paper insulated cables for electricity supply

I.E.E. Wiring Regulations particularly applicable to this chapter are:

Section   537

7 Lighting

Introduction

Illumination and the design of lighting layouts is a subject on its own. There are books dealing comprehensively with it and it is not proposed to condense the matter into a single chapter here, but once a lighting layout has been arrived at, it is necessary to design the circuits, wiring and protection for it; this is an aspect of lighting design which tends to be overlooked in books on illumination and which we propose to discuss in this chapter. The electrical requirements of a lighting system depend to a considerable extent on the kind of lamps used and we shall describe the different available types in turn.

Incandescent lamps

These are also known as tungsten lamps, and are in fact the ordinary bulbs still most commonly used in homes. They consist of a thin filament of tungsten inside a glass bulb. When a current is passed through the filament, heat is produced and the temperature of the filament rises. The filament is designed so that it reaches a temperature at which it generates light energy as well as heat, which means that the filament glows or is incandescent and hence the lamp is called an incandescent one. The higher the temperature of the wire, the more efficient is the conversion of electrical energy into light energy, but if the temperature becomes too high the wire melts and breaks. Tungsten has a melting point of 3382°C, and most modern lamps have filaments running at about 2800°C, although some special lamps may run at 3220°C.

The colour of the light produced depends on the temperature, becoming more white as the temperature rises. At 2800°C, it is rather yellow, but as no material is known which can be operated at a higher temperature than tungsten, lamps of this type cannot be made to give a daylight colour.

To prevent the filament from oxidizing, all the air must be evacuated from the bulb, and the early lamps were of the vacuum type. It was found that in a vacuum tungsten evaporated and blackened the inside of the bulb. This problem has been solved by filling the bulb with an inert gas at a pressure such that when the bulb is hot the pressure rises to about atmospheric pressure. The gas used is generally a mixture consisting of 93% argon and 7% nitrogen. Unfortunately, the gas conducts heat from the filament to the bulb, thus lowering the temperature of the filament and reducing the efficiency and the light output of the lamp. To overcome this effect as much as possible, the filament must be wound to take up as little space as possible. It is for this reason that gas filled lamps have the filament either single coiled or coiled coil.

The connections of the filament are brought out to the lamp cap. This is the end of the lamp which fits into the lampholder when the lamp is put into a light fitting. Lamp caps are shown in Fig. 68. The commonest is the two-pin bayonet cap (Fig. 68a) which is standard in the United Kingdom for all sizes up to and including 150 W. It can be inserted into the lampholder either way round and for applications where the lamp must be fixed in one position only a three-pin bayonet cap is available. Clearly a special lampholder is needed for it. Fig. 68c shows the Edison screw cap, in which the screw thread forms one of the terminals. The Edison screw cap is made in five sizes covering all types of lamps from street lamps to flash bulb lamps, but the only common sizes used in this country are the Goliath (usually abbreviated to G.E.S.), small (S.E.S.) and miniature (M.E.S.). Bulbs of 200 W and over use the G.E.S. cap.

It will be noticed that the Edison screw affords slightly greater risk of accidental contact with the terminal when one is putting a bulb into a lampholder. The screw thread is the neutral which should never be at line

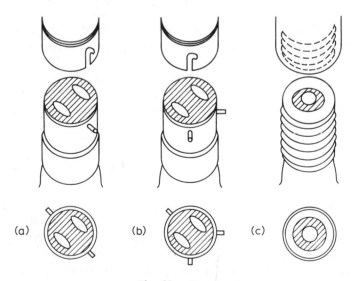

(a)   (b)   (c)

Fig. 68   Lamp caps

voltage, and it is possible to hold the bulb without touching the thread, so that
there are two reasons why there should not be any danger. Nevertheless, if the
circuit wiring is of the wrong polarity, which is not an entirely unknown fault to
occur, and at the same time an inexperienced person is clumsy with the bulb, he
could get a shock. This may be a remote possibility but is probably the reason
why the United Kingdom adopted the bayonet cap as the standard up to 150 W,
which is the largest bulb likely to be used in domestic premises. Some
continental countries use the Edison screw for all sizes.

The characteristics of incandescent lamps are shown in Fig. 69. The graph
shows how very sensitive the life is to change of voltage. For example an increase
of 2½% above normal voltage increases the efficiency by 2½% and the light
output by 7% but reduces the life by 20%. The graph also shows that for
voltages below normal the light output falls more rapidly than the voltage, which
is something to be borne in mind when one considers the voltage drop that can
be accepted in the circuit wiring to the lamps.

The current taken by an incandescent lamp is purely resistive. Because of this,
there is no adverse effect on the circuit power factor and no special
consideration has to be given to the switches used. The situation is different for
most of the other types of lamp.

The heat given off by incandescent lamps, especially the larger sizes, must be
taken into account in both the design of incandescent light fittings and in the

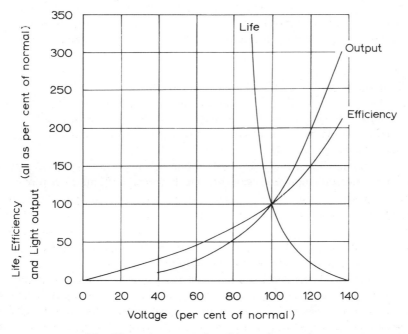

Fig. 69   Characteristics of incandescent lamps

selection of fittings for a lighting scheme. The fittings must allow enough natural ventilation to keep the normal working temperature of the fitting and wiring reasonably low. This is particularly important where fittings are made of plastics with softening temperatures in the region of 60°C. In such cases, the maximum size of bulb which may be used in a fitting is limited not by the physical dimensions of the fitting but by the heat generated.

When incandescent lamps are used for stage lighting or for special effects they can be dimmed by the insertion of resistance in series with them. This affords a simple way of gradual and continuous dimming.

Voltage control units with solid state electronics can be made sufficiently compact to be incorporated in a light switch. Combined switch and dimmer units of this type are used where variable light effects are required in places like hotels and luxury housing. The electronic circuits need additional protection and the units therefore include their own fuses. They must also include a radio interference suppressor.

Fluorescent lamps

The action of a fluorescent lamp depends on the discharge of a current through a gas or a vapour at a low pressure. If a tube containing a vapour has an electrode at each end, a current will flow through the vapour provided electrons are emitted by one electrode (the cathode) and collected by the other (the anode). Electrons will be emitted if the potential gradient from anode to cathode is great enough; the potential difference required to cause emission decreases as the temperature of the cathode increases, and therefore lamps designed to operate at normal mains voltage have cathodes which are heated to a dull red heat. They are known as 'hot cathode' lamps.

Even when the cathode is heated, a voltage has to be applied between the electrodes to start the discharge, and the minimum voltage needed is known as the striking voltage. After the discharge has started a voltage is still needed between the electrodes to maintain the discharge, but the maintaining voltage is less than the striking voltage.

A current flowing through a gas or vapour at low pressure causes the gas or vapour to emit radiation at wavelengths which depend both on the nature of the vapour and on its pressure. An incandescent lamp gives out light energy at all wavelengths, whereas a discharge lamp gives it out at certain discrete wavelengths only. The wavelength of the radiation emitted may be in the visible spectrum or above it or below it, and one of the functions of the lamp is to convert all the primary radiation into useful visible radiation.

A fluorescent lamp consists of a long glass tube containing a mixture of mercury vapour and argon gas at a pressure of 2 to 5 mm mercury. When the lamp is cold, the mercury is in the form of small globules on the tube surface, and the argon is needed to start the discharge. As soon as the discharge starts the temperature rises sufficiently to vaporize the mercury which then takes over the conduction of practically the whole current. At either end of the tube, there is an electrode made of a tungsten filament coated with an alkaline earth metal having suitable electron emission properties. Each electrode acts as cathode and anode on alternate half cycles of the a.c. supply. Anode plates in the form of metal fins are provided round each electrode to assist it in collecting electrons during the half cycle in which it acts as anode. The inside of the tube is coated with a fluorescent powder. A fluorescent material is one which has the property of absorbing radiation at one wavelength and emitting radiation over a band of wavelengths in another region of the spectrum. It emits radiation only while receiving it; a material which continues to emit after the incident radiation has ceased is called phosphorescent and it is an unfortunate confusion that the fluorescent materials used in commercial lamp manufacture are commonly called phosphors.

Thus in the fluorescent lamp the radiation emitted by the current discharge through the mercury vapour is absorbed by the fluorescent coating which then emits a different radiation. The fluorescent coating is most susceptible to excitation by ultraviolet radiation, and it is the need to have the radiation from the mercury in this region that determines the operating pressure. The secondary radiation emitted by the coating is in the visible spectrum and its colour depends on the material used for the coating. So many fluorescent materials are now known that it is possible to obtain almost any colour, including an almost exact reproduction of daylight.

The circuit needed to operate such a lamp is shown in Fig. 70. As we have explained, the voltage required to maintain the discharge is less than that

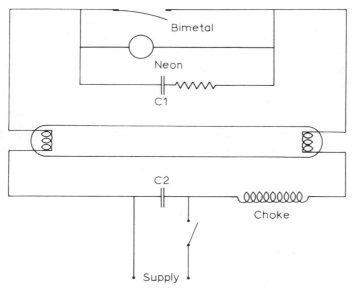

Fig. 70   Fluorescent lamp circuit

required to start it, and therefore once the discharge has started the voltage across the lamp must be reduced. If it were not reduced the current through the lamp would go on increasing until the lamp was destroyed. The necessary reduction is achieved by a series ballast which takes the form of an inductance or choke. Initially, when there is no discharge through the lamp, the entire voltage of the mains is applied across the electrodes. As soon as the discharge is established, current flows through the lamp and choke in series, a potential difference is developed across the choke, and the voltage across the electrodes is reduced by the voltage across the choke.

The circuit also includes a starter switch which consists of a small neon glow lamp and a bi-metal strip. When the lamp is cold, the bi-metal switch is open so that the whole of the mains voltage appears across the neon glow lamp which discharges. The heat of the discharge heats the bi-metal until the contacts on the end of the bi-metal close. There is now a circuit formed by both electrodes of the main lamp, the bi-metal and the choke. A small current flows through this circuit and heats the electrodes. However, the bi-metal short circuits the neon, which ceases to glow and, therefore, to heat the bi-metal. In consequence, the bi-metal cools and after a time it opens again. This interrupts the circuit which, being highly inductive, responds with a sharp voltage rise across the switch. Since the switch is in parallel with the lamp the voltage rise is also applied across the lamp and is sufficient to start the discharge. The choke now takes the normal current and reduces the voltage across both lamp and switch. This reduced voltage is not enough to start another glow at the switch, but if for any reason the main discharge fails to start, then mains voltage again appears at the glow lamp and the sequence starts again. This happens if the choke is faulty and reduces the voltage across the lamp below that required to maintain the discharge. When the lamp has started the electrodes are kept hot by the current through the lamp which also flows through the electrodes.

The starting sequence described takes a few seconds. This delay can be avoided with the instant start circuit shown in Fig. 71. The electrodes are supplied from low voltage secondaries of a transformer, and carry the full heating current continuously. A metallic strip runs the whole length of the tube and close to it, and is connected to earth. When the lamp is switched on a capacitative current flows from the electrodes to the earthed strip and is just sufficient to ionize the gas in the tube and thus enables an arc to strike from end to end. As soon as the arc has struck, the choke reduces the voltage across the lamp to its normal operating value and the current in the heaters also reduces as a result of this. Although it is possible to use ordinary lamps with quick start circuits, it is not wise to do so because the heaters carry a higher current than when they are used with switch start circuits. For this reason, only tubes designed for quick start should be used in fittings having quick start circuits.

Fig. 70 shows a small capacitor across the switch contacts. The capacitor is inserted to suppress radio interference. Figs. 70 and 71 both show a larger

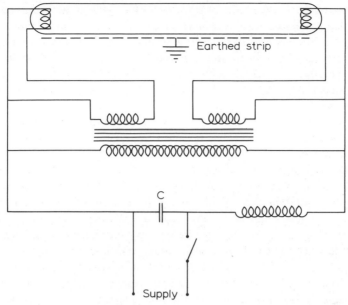

Fig. 71   Quick start circuit

capacitor across the entire circuit. This one is included to correct the power factor, which would otherwise be unacceptably low because of the inductive choke.

The capacitors, switch and choke are usually housed within the light fitting which holds the lamp. It is important that the fittings used are suitable for the lamps it is intended to employ. The current taken by a fitting with fluorescent lamps is inductive with a power factor of about 0.8. The switches controlling fluorescent lights, therefore, have to break inductive circuits and must be capable of withstanding the voltage rise which occurs when an inductive circuit is broken. The voltage rise can cause arcing across the switch contacts, and if this is serious enough and occurs often enough it will destroy the switch. Most modern switches are capable of breaking an inductive current of the value at which the switch is rated, but the older switches designed only for incandescent lights had to be de-rated when they were used on circuits serving fluorescent lights.

There are some occasions when the switchgear for the fitting, that is to say the capacitors, switch and choke, is mounted remotely from the fitting. One case in the author's experience was in a swimming pool where the lamps were mounted behind a pelmet round the perimeter of the hall. The architectural design was such that there was no room for ordinary fittings containing the gear, and the lamps were simply clipped to the pelmet. Each end of each lamp was plugged into a two pin socket from which wires ran to a remote cupboard. Inside the cupboard there was a series of trays each of which contained the control gear for one lamp. Another case was in an air conditioned computer room. Much of the heat lost from a fluorescent light is generated in the control gear rather than in the lamp, and in order to keep the heat load on the air conditioning plant as low as possible, it was decided to keep the control gear out of the air conditioned room. The method used for doing this was similar to that in the previous example.

Fluorescent lamps have a lower surface brightness than incandescent ones and have a higher efficiency in terms of light energy given out per unit of electrical energy consumed.

Fluorescent lamps cannot be dimmed simply by the insertion of a resistance to reduce the current through the lamp. If a reduced current flows through the tube, it does not heat the cathode enough to bring about emission of electrons. Successful dimming requires that the electrodes be permanently heated while the tube current is varied, and this can be done by supplying the electrodes with a separate heating current from a low voltage transformer. A suitable circuit is shown in Fig. 72 and the similarity with the quick start circuit of Fig. 71 can be noted. The transformer provides a permanent heating current which is independent of the current through the tube and the latter current is varied by the resistor. In this way, variation of the tube current does not cut off electron emission from the cathode.

The tube current can also be varied by a thyristor. The arrangement is similar to Fig. 72, but a thyristor circuit takes the place of the resistor. While permanent heating current is circulated to the electrodes, the thyristor controls the proportion of each cycle of the alternating supply during which striking voltage is applied to the tube. This determines the length of time in each cycle during which current flows through the tube, and hence the light output.

Mercury discharge lamps

Mercury discharge lamps operate on the same principle of discharge as fluorescent lamps. The difference is that they do not have a fluorescent coating and the light given out by the lamp is the primary radiation from the discharge itself. The discharge must, therefore, produce radiation in the visible spectrum and not in the ultraviolet range, and to get it into the visible region these lamps work at high pressures. There are several different versions of discharge lamps.

The basic mercury discharge lamp consists of a glass envelope with an electrode at each end, the electrode being similar to that in the fluorescent lamp. There is also a starting electrode near one of the main electrodes and this is connected to the other main electrode through a high resistance. The first mercury lamps were designated MA/V for those which worked vertically and MA/H for those which worked horizontally. They operated at a pressure of 1 atmosphere. The old MA/V lamp is shown in Fig. 73.

A modern MB/U lamp is shown in Fig. 74. It has two main electrodes with a

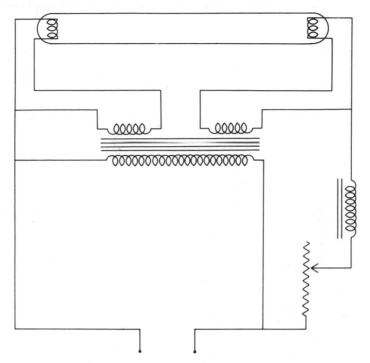

Fig. 72 Fluorescent dimming circuit

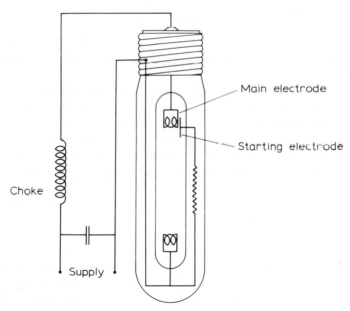

Main electrode

Starting electrode

Choke

Supply

Fig. 73 MA/V mercury lamp

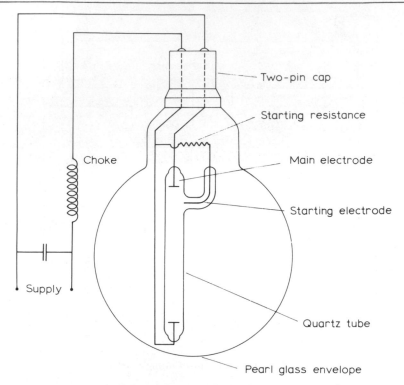

Fig. 74   MB/U mercury lamp

starting electrode near the upper one connected to the lower one through a high resistance.

The electrodes are contained in a quartz envelope which can resist the heat of the arc. This envelope becomes hot when the lamp is on and it is therefore surrounded by a second outer glass envelope, the space between the two envelopes being evacuated. The operating pressure inside the inner envelope is 5 to 10 atmospheres. The outer envelope is a pearl glass bulb about the same size as a 100 W tungsten lamp. It has a three-pin bayonet cap and this makes it impossible to put the lamp into an ordinary domestic lampholder. The inner surface of the outer tube is coated with a phosphor. This increases the efficiency slightly but its chief function is to improve the colour rendering.

When the lamp is switched on the voltage is sufficient only to start a small discharge between the starting electrode and the upper main electrode. The small discharge makes the argon in the tube conductive and this makes it possible for a discharge to start between the main electrodes. This discharge bypasses the starting resistance, warms the tube and evaporates the mercury. As it evaporates, the mercury takes over conduction of the main discharge. After two or three minutes, steady conditions are reached and the mercury gives out a light the colour of which is modified as it passes the phosphor on the outer tube. The current through the discharge is limited by the ballast choke.

When it is switched off, the lamp will not restart until it has cooled and the pressure has dropped again. This takes two to three minutes, but it will not come to any harm if it is left switched on while it is cooling down. In other words, if the lamp is switched off and switched on again immediately, it will not relight immediately but will not be damaged.

It will be noticed that there is a capacitor for power factor correction included in the circuit. The capacitor and choke must be contained either in the light fitting or in a casing conveniently near the fitting. These lamps are still sometimes used for street lighting, and the control gear can be accommodated inside the column on which the light is supported. If it is intended to do this, the column manufacturer's specification should be checked to make sure that the column is large enough to hold the gear.

A variation on this lamp is one which has an additional tungsten filament inside the outer envelope. The tungsten filament is connected in series with the discharge tube and acts as a ballast. Consequently, there is no need for a choke. The light from the tungsten filament also adds colour correction to the light from the discharge so that there is no need to have a phosphor coating on the outer tube. The lamp is known as the MBT/U. Because the tungsten filament used as a ballast is not inductive, no power factor correction capacitor is required. There is therefore no external gear at all for this lamp.

The ME is a lamp which operates at a working pressure of about 20 atm and at a very high temperature. This necessitates a special construction of the electrodes and the use of quartz instead of glass for the enclosure. At powers above 1000 W a.c. operation leads to pitting of the electrode surfaces and d.c. operation is preferable. In the MEC lamp cadmium or cadmium and zinc are amalgamated with the mercury, and the amalgam produces a colour very close to white. The light output is very high and these lamps are used for television and film studio lighting. ME lamps are used for optical projection, for which they are suitable because of the small physical dimensions in relation to the amount of light produced.

The MD lamp operates at pressures of 80 to 100 atm This makes it so hot that it has to be water cooled. Consequently it can only be used in optical equipment which incorporates a water pumping system; even then it has a life of only about 100 hours.

Sodium discharge lamps

Sodium discharge lamps have a similar action to mercury lamps, but the filling used is sodium instead of mercury. Ordinary sodium lamps work at low pressure and their luminance is low. They therefore have to be very long to give a good light output, and in order to keep the overall length down to a reasonable size the lamp tube is bent into the shape of a 'U'. The resulting construction is shown in Fig. 75.

To withstand the sodium vapour the inner tube is made of ordinary glass with a thin coating of special glass fused onto its inner surfaces. The inner tube is enclosed in a double walled vacuum flask. Each electrode consists of a coated spiral whose ends are twisted together; there is no flow of heating current through the electrode as there is in a mercury lamp. Neon is contained within the inner tube with the sodium and starts the discharge. When the lamp is cold the sodium condenses and exists as small globules along the length of the tube. It is important that they should be fairly evenly distributed along the tube, and, therefore, the lamp must be kept nearly horizontal. At the same time, the sodium must not be allowed to condense on the electrodes. To satisfy this requirement fittings for sodium lamps are arranged to hold the lamp tilted slightly above the horizontal.

The operating pressure is very low, being in the region of 1 mm mercury, although the vapour pressure of the sodium alone is of the order of 0.001 mm mercury, the rest being due to the neon. To start an arc through the neon when the lamp is cold requires a voltage higher than ordinary mains voltage (about 450 V). The necessary striking voltage is obtained from an auto-transformer which is specially designed to have poor regulation, that is to say the voltage when current flows drops greatly below the no load voltage. Consequently, as

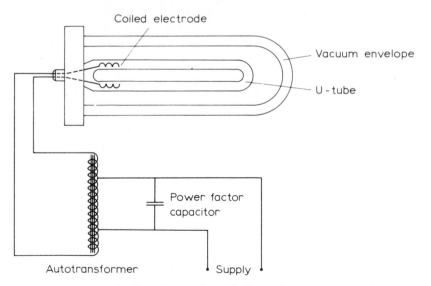

Fig. 75   Sodium discharge lamp

soon as the discharge starts the voltage drops to that required to keep the discharge going. The transformer thus performs the functions of the ballast and no separate choke is required. A capacitor for power factor correction is however needed.

The discharge which starts in the neon is of a red colour. This warms the tube and gradually vaporizes the sodium. After about twenty minutes, the sodium is fully vaporized and gives its characteristic yellow colour. The sodium discharge lamp is the most efficient means so far known of converting electrical into light energy, but because of its peculiar colour the low pressure sodium lamp is limited to street lighting and similar applications. The control gear, consisting of auto-transformer and power factor capacitor, is usually accommodated within the column which supports the fitting. Alternatively, it can be housed within the fitting, and this is done in floodlights and other fittings intended for mounting at low level.

The solarcolour is a high pressure sodium lamp and gives a rather sunny yellow light. It is suitable for factories and warehouses as well as for street lighting and other outdoor applications. The pressure of the sodium when the lamp is fully warmed up is in the order of 250 mm mercury. The lamp runs at a temperature of 1300°C and to withstand the corrosive properties of sodium at this temperature crystalline alumina is used in the manufacture of the tube.

Like the low pressure sodium lamp, the solarcolour lamp does not have heated cathodes or auxiliary electrodes, but starts cold with a high voltage pulse. A typical circuit is shown in Fig. 76. A thermal starting switch is generally

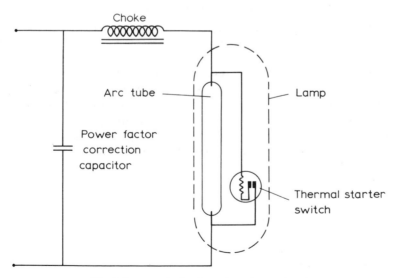

Fig. 76   Typical solarcolour lamp circuit

incorporated inside the lamp itself. When the lamp is off, it is cold and the contacts of the bi-metal are closed. When the lamp is switched on the heater coil warms up and ultimately opens the bi-metal switch. The opening of the switch breaks the inductive circuit through the choke and produces the high voltage needed to strike the arc. The striking voltage is between 1000 and 4000 V. If, however, the switch opens at a point in the alternating supply cycle at which the current is either zero or very low, the induced voltage is not enough to strike the arc. The heater coil must then cool down to allow the bi-metal contacts to close again before the starting sequence can repeat. The normal switching on time is about 1 min and the cooling down time about 2 min. If the switch opens at a point of low current on several successive occasions, it may take the lamp many minutes to start but the probability of this happening is very low The manufacturers can provide an external starting device which can be included in the lamp circuit if delayed starting is totally unacceptable.

Solarcolour lamps can work either vertically or horizontally. They may be mounted at any angle below 20° above the horizontal. The tube does not have to be as long as the low pressure sodium tube and the construction and shape are similar to those of the mercury discharge lamp. Indeed, the first solarcolour lamps made were designed to be capable of operating in existing mercury lamp installations with the existing control gear. The later and higher output solarcolour lamps are intended for new installations with control gear specially designed for them.

Cold cathode lamps

These include both neon advertising lights and lamps for illumination used in large stores, cinemas and similar areas.

If a sufficiently large potential difference is applied between the electrodes of a discharge tube, the arc can be struck and maintained without any heating of the cathode. With a hot cathode, the volt drop across the electrode is small, but with a cold cathode, it is higher. A long tube helps to keep a larger part of the total applied voltage drop across the arc and a smaller part across the electrode, and cold cathode lamps are, therefore, made longer than the hot cathode lamps described previously. The greater length, in fact, increases the efficiency and hence also the light output. It is because of this that they are used in stores and cinemas and under the projecting canopies which are now so popular at the entrances to commercial buildings, including hotels.

The high voltage required is provided by transformers, and in order to keep the amount of high voltage wiring to a minimum the lights are supplied with transformers in self-contained units suitable for direct connection to the mains. A common arrangement has three 9 ft tubes physically parallel to each other, but electrically connected in series. The circuit diagram of such a fitting is shown in Fig. 77. The two transformers and all the high voltage wiring are sealed within the fitting and are not accessible.

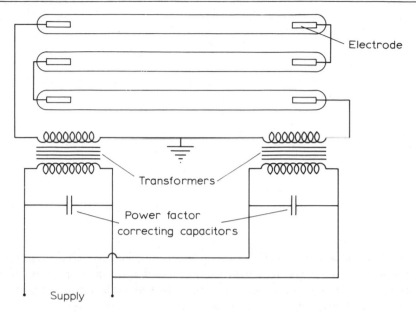

Fig. 77   Cold cathode lamps

The two primaries are in parallel across the mains supply, and each has a power factor correcting capacitor. The high voltage windings each have one end connected to earth and are so arranged that the voltage between their two other ends is double the maximum voltage to earth. The transformers give a voltage high enough to strike the arc. As in the case of the transformers used with the hot cathode sodium lamps, they are designed to have poor regulation so that once the arc has been struck the voltage drops to that required to maintain the arc. For a typical three tube fitting the striking voltage is 3600 V and the running voltage 2000 V.

The great advantage of the cold cathode tube is its very long life, which is in the region of 15 000 hours as against about 5000 hours for an ordinary fluorescent tube. It also maintains its output better throughout its life, and starts instantly. Its life is not reduced by frequent switching. When the tube is used for lighting it is filled with mercury vapour and has a suitable fluorescent coating on its inside surface. As with ordinary fluorescent tubes a variety of colours is available.

The length of cold cathode tubes makes them suitable for bending into special shapes. This makes them useful where special decorative effects are wanted and also, of course, for advertising purposes. For advertising use many different colours can be obtained by the use of different types of glass, which may be coloured or have fluorescent coatings. The earliest gas used was neon, and 'neon light' has stuck as the popular generic name for all advertising lights of this type.

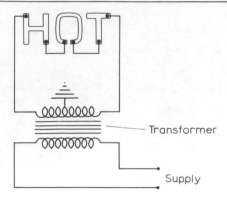

Fig. 78   Decorative cold cathode lamps

The tubes can be bent into the shape of letters or other symbols and successive tubes can be connected in series, an example of such an arrangement being shown in Fig. 78. For letters like H or T the tube must be bent back on itself.

The voltage required to start the tubes depends on their length, the type of gas and its pressure and on the design of the electrodes. The manufacturers normally quote the running voltage per metre of tube and the voltage drop at each pair of electrodes, from which the designer can calculate the voltage which he has to provide from the secondary of the transformer. A magnetic shunt is provided inside the transformer to enable the tube current to be adjusted when the sign is commissioned. This is not a task to be undertaken lightly, because of the very high voltages involved.

Stringent safety regulations for high voltage advertising signs have been drawn up by the Institution of Electrical Engineers, and recommendations are also given in BS.559:1959. No part of the installations may have a voltage to earth greater than 5000 V, and if the centre point of the secondary side is earthed the voltage available between the opposite ends of the series of tubes is 10 000 V. This limits the length of tubes which can be served from one transformer. If greater lengths are required, the sign must be split into several lengths each having its own transformer. The transformers are normally housed inside weatherproof steel containers which are fitted on the outside of the building. A transformer rated at more than 250 VA must be supplied on a separate sub-circuit not serving any other equipment or any other transformer. If the transformers are smaller than this, up to four of them may be put on a common sub-circuit, but the total rating of one circuit may not exceed 1000 VA. Particular care must be taken with the high voltage wiring which should be lead covered armoured cable. It must be restrained from swaying in the wind because this would strain both the conductor and the insulation.

A typical circuit is shown in Fig. 79. A lockable switch is provided so that

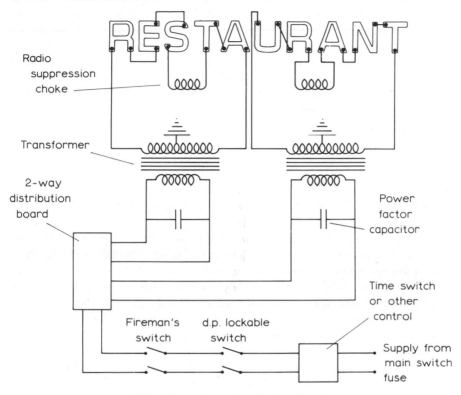

Fig. 79 Circuit for display lighting

anyone doing maintenance on the installation can be sure that no other person can inadvertently turn the supply on. The fireman's switch is mounted outside the building in a prominent position but out of reach of the public. It is for the use of the fire brigade, who want to turn off all high voltage sources before spraying water anywhere near them A choke is connected in each high voltage circuit to suppress radio interference.

Flicker

With the exception of the incandescent ones, all the lamps we have described go on and off 100 times a second when working on a 50 Hz supply. This is too rapid a flicker to be noticeable to the eye, but it can produce a stroboscopic effect on rotating machinery, which may appear either to be stationary or to be moving at a much slower speed than it really is. Such an illusion can obviously be a source of danger and should be avoided. If adjacent lamps go on and off at different times, that is to say if they are out of step with each other, the effect is reduced, and it is possible to obtain this breaking of step with twin lamp fittings.

The fitting is made with one choke for each lamp but only one capacitor. The capacitor is arranged in series with only one of the lamps, and the chokes and capacitor are so sized that although the combined current through both lamps has the usual power factor of about 0.8 the current through one lamp leads the voltage while the current through the other lags. The two lamp currents are neither in step nor 180° out of phase and, therefore, extinguish and re-light at different instants. Fittings of this kind are known as lead/lag fittings and are often specified for workshops and factories.

Circuits and controls

The way in which light fittings are connected in circuits has been described in Chapter 5. We showed in that chapter how each light or each group of lights can be controlled by a switch. The switch would be one of the types described in Chapter 1. There are, however, other methods of control possible and they are particularly important for street lighting and other external lights for which some form of automatic control is almost essential.

The first method and one still widely used is by means of a time switch. Time switches are made with an enormous variety of dials, and a dial for almost any conceivable application can be found from some manufacturer's standard range of products, but in exceptional cases special dials can be made to order. For street or external lights, the usual requirement is for the lamps to be turned on at sunset and off either at midnight or at sunrise. The times of sunset and sunrise vary throughout the year, and solar compensating dials are made for time switches to follow the annual variation in sunset and sunrise. As the variation depends on latitude, different dials are made for different zones of latitude.

In the United Kingdom street lighting is the responsibility of the local authority. The usual arrangement is for the Electricity Board to supply each column directly off its main running through the street, or if necessary off a main laid specially for the lights. A time switch is fixed inside each column, and switches the current to the fitting at the top of the column. The supply is not metered because one knows the total hours in a year during which a solar dial switch is on and also the rating of the lamp, and can thus calculate the number of units of electricity consumed in a year. The column also contains a service head fuse, on the supply side of the time switch.

A new housing development may often include estate roads which are constructed by the developer, but which will ultimately be taken over by the local authority. The same system of road lighting can be adopted, but it is advisable to agree the details with the street lighting department of the local authority beforehand. The system can also be used where the roads are to remain private, but the agreement of the Electricity Board must be obtained in advance because the method of supply may affect the tariff they will want to apply.

The design engineer should make sure he selects and specifies columns which have a space inside them large enough to contain the time switch and fuse; there are light duty columns which do not have a wide enough base.

If the external lights are not too far from buildings, they can be served by sub-circuits from distribution boards within the buildings. The cable sizes should be checked for voltage drop, and may need to be greater than ordinary lighting circuits. For running underground PVC served MICC cable is probably the most suitable, although PVCSWAPVC is also sometimes used. A circuit to several lights is taken through a time switch and the time switch can be conveniently mounted next to the distribution board. It is also possible to obtain a distribution board which has enough space in it for the time switch to be mounted inside the board. Internal lights may also be controlled by a time switch, and this may be a convenient way of switching, say, corridor lights in an office block.

When lights, whether internal or external, are supplied through a time switch served from a distribution board, the supply is of course part of the metered supply to the building. Whether this is more or less desirable than an unmetered but timed supply depends on the agreement between the consumer and the supply authority.

The other method of automatic control of lights is the use of photo-electric cells. In this method, a photo-electric cell is arranged to make a circuit when the illumination falls below a set value and to break the circuit when the illumination rises. The advantage over a time switch is that the control takes account of weather conditions. There are summer evenings which, because of storm clouds, are almost as dark as winter evenings, but a time switch cannot distinguish between a cloudy day and a clear one. Furthermore, in the United Kingdom at least, a time switch must usually be reset twice a year because of Summer Time, whereas a photo-electric cell needs no alteration.

When photo-electric cells are used, the lighting circuits can be arranged in the same way as when time switches are used. There can be one cell to each light, or there may be one cell for a circuit on which there are several lights. In the latter case, if the circuit is switched directly by the cell, the number of lights on the circuit is limited by the current which the cell can switch. The maximum current which a given cell can switch may be lower if the current is inductive than if it is resistive, so that a cell may be able to switch fewer fluorescent lights than incandescent ones. Nevertheless, a small cell can be used to control many lights, or even several different circuits, if it is used with a relay. The cell is placed in the operating coil circuit of the relay and the relay main contacts switch the lights. A multipole relay makes it possible for one photo-electric cell to control several lighting circuits.

The position of the photo-electric cell requires a little thought. It must be close to the area illuminated by the lights it controls, in order to react to the daylight in that area. At the same time, it must not receive direct light from any of the lamps it controls; if it did, as soon as they came on, it would switch them off again, so that they would flicker on and off continuously. Nor must it receive direct light from any other lamps which may be turned on at night, otherwise switching on one set of lights will immediately switch off the other set. In general, the illumination on the photo-cell during the hours of darkness when the lights are on must be less than the level at which the cell switches the lights. Most outdoor artificial lighting produces illumination levels much lower than daylight and a photo-cell set at the minimum acceptable daylight level will not turn the lights off until daylight is again adequate, but there are areas, such as the forecourts of garages, supermarkets and cinemas, where the artificial lighting approaches full daylight level, and in these cases it is certainly difficult, and may even be impossible, to design a photo-electric scheme of control. For ordinary street lighting, a cell placed in the top of a column above the lamp fitting will respond adequately to natural light falling on the road and can be shielded from the light of the lamp.

The number of hours in a year during which a photo-electric cell will be in the 'on' position cannot be known exactly. The supply authority may, therefore be unwilling to agree to the use of an unmetered supply. Nevertheless, several local authorities are using photo-electric cells to control street lighting and have negotiated a suitable method of payment with the Area Electricity Board.

Emergency lights

Many buildings must have some form of emergency lighting to come on if the electric supply to the ordinary lights fails. One of the possible causes of failure is a breakdown in the supply authority's service to the building and, therefore, the emergency supply must be independent of the service into the building. Gas is still sometimes used for emergency lighting and is both effective and efficient. Its use is nevertheless decreasing and in any case a discussion of gas lighting is outside the subject matter of this book.

Electric lighting for emergency use can be provided if the building has a standby generator. A generator can be installed to take over the entire supply to a building, so that no special provision for emergency lights need be made, but for economy the standby generator is often rated at less than the ordinary mains service to the building. The distribution then has to be arranged so that only a part of the service within the building is fed by the generator, and only a few of the lights should be included in this part. There is no need for full lighting under emergency conditions, and lighting in the main corridors and staircases is usually enough.

Emergency supplies are of particular importance in hospitals and no new hospital should be built without a standby generator, but buildings like schools, offices, theatres and blocks of flats seldom justify the expense. For these buildings emergency lighting is almost invariably provided by a battery system.

Emergency lights are fitted throughout the building. They come on only when the mains fail and cannot be used while the mains are healthy. They are not intended to give full illumination, but only to provide sufficient light for people to make their way out of the building safely. One light on each landing and perhaps one in the centre of any particularly long corridor should be perfectly adequate. These lights work on low voltage d.c. and are fed from a battery. A trickle charger permanently connected to the mains ensures that the battery is always fully charged. The lights are wired from the battery through a relay, the contacts of which are closed when the coil is de-energized. The coil is fed from the mains and as long as the mains are on, the contacts are held open. Thus as long as the main supply is healthy the battery lighting circuit is kept open, but immediately the mains fail the relay contacts close and the emergency lights come on. The circuit diagram is given in Fig. 80.

There are battery chargers and relays purpose-made for this kind of application. The charger must be left permanently switched on, and contains the

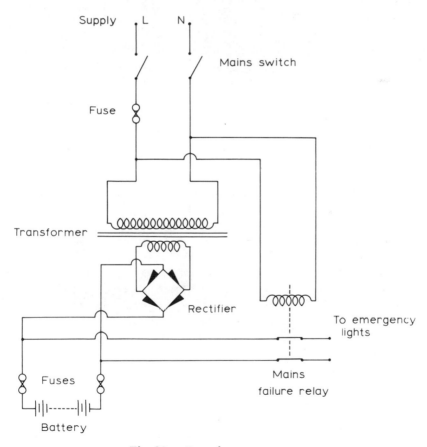

Fig. 80   Central emergency system

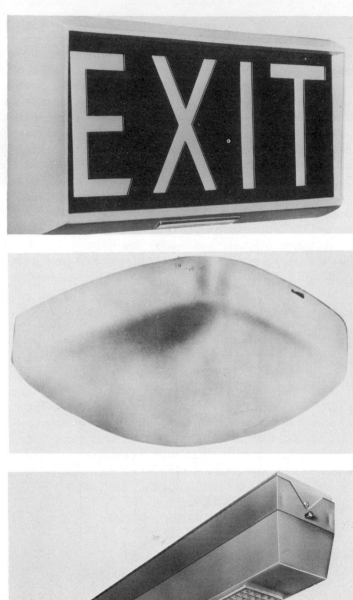

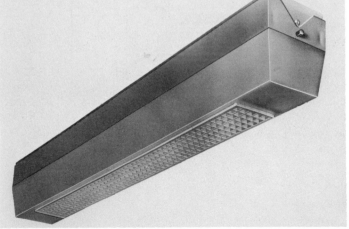

Fig. 81 Emergency lights (*Courtesy of* Bardic Systems Ltd.)

relays necessary to stop the charging current when the battery is at full charge. It can be supplied by a sub-circuit from any convenient distribution board, but there should be no other outlets on the same sub-circuit. Alternatively, it can be fed directly from a switch fuse at the main intake.

Because the emergency lights work at a low voltage, the voltage drop in the cable to them can become considerable and may present something of a problem. Whereas a 60 W bulb on a 240 V supply takes 0.25 A, a 24 W bulb on a 24 V supply takes 1.0 A and the voltage drop in a cable of a given size is four times as great. At the same time, a drop in a cable of 2.4 V in 240 may reduce the light output by perhaps 2% but the same drop of 2.4 V in a 24 V system is proportionately ten times as great and could reduce the light output by a fifth or a quarter. Low voltage cables must, therefore, be adequately sized. It is in any case inadvisable to design an emergency system for less than 48 V d.c., which is a convenient standard battery output voltage. Even with a 48 V system and ample cable sizes, there must obviously be a limit to the number of lights which can be served from one battery and to the distance the furthest light can be from the battery. A large building may, therefore, need several separate battery systems. Recent legislation has made it more essential to provide lights to mark fire escape routes from buildings. The lights used for this are of very low wattage and, consequently, the voltage drop problems are somewhat eased.

The emergency lights themselves are ordinary fittings which take a low voltage d.c. bulb. There is another method of emergency lighting which makes use of special fittings each of which contains its own battery, charger and relay. The fitting effectively houses a complete low voltage system just large enough to operate one light. The use of such fittings makes it unnecessary to run a low voltage circuit throughout the building. With this system, the emergency fittings are put in the most suitable places for emergency illumination and are fed from any convenient lighting circuit. In some cases, it may be convenient to have two or three emergency lights on a circuit of their own and in other cases, it may be convenient to have an emergency light included in one of the normal lighting circuits.

Such self-contained emergency light fittings are made in a variety of shapes and with a variety of lamps. They may give incandescent or fluorescent light, and can take the form of illuminated signs. An emergency light with the word 'EXIT' has an obvious application. Examples of emergency lights are illustrated in Fig. 81. An important advantage of these self-contained fittings is that they are sealed for life and require no maintenance.

Flameproof fittings

The principles on which flameproof accessories are designed are explained in Chapter 1. Most types of light fitting are also available in flameproof versions. When these are specified it is important to make sure that the ones selected are suitable for the group into which the vapour causing the hazard falls. The majority of flammable gases falls into Group II and most flameproof equipment is designed for this group, but if the danger is due to a Group III gas a Group II fitting will not give adequate safety.

Standards relevant to this chapter are:

BS 52	Bayonet lamp caps, lamp holders and adaptors
BS 559	Electric signs and high voltage luminous discharge tube installations
BS 889	Flameproof electric lighting fitting
BS 1875	Bi-pin lamp caps and lamp holders for fluorescent lamps
BS 2560	Exit signs
BS 4533	Electric luminaires (lighting fittings)
BS 5042	Lamp holders and starter holders

8 Power

The majority of outlets to be used for power services are socket outlets of the types described in Chapter 1. The various ways in which power circuits can be arranged have been described in Chapter 5. It is usually found that as soon as more than two or three socket outlets are to be supplied, it is more economical to serve them from ring circuits than from radial ones. On ring circuits, only 13 A socket outlets should be used in order to ensure that the plug must be of the fused type and that the appliance and flexible cable to it do not rely on the ring circuit fuse for protection. There is little more one can add about socket outlets and it remains in this chapter to say something about connecting fixed appliances and larger equipment.

Fixed appliances of small ratings, by which we mean up to 3 kW, can be served through fused spur units from the ring mains serving the socket outlets in the same area as the fixed appliance. A 3 kW electric fire is one example of a fixed appliance which might be supplied in this way. If the socket outlets in the area are on radial rather than ring circuits then each fixed appliance must have a separate radial circuit of its own. It is often convenient to supply equipment such as motorized valves on hot water heating systems, roof mounted extract fans in kitchens, tubular heaters in tank rooms and so on by a separate circuit for each item or group of items. There is nothing wrong technically with supplying them by a spur from an adjacent general purpose ring main provided the permanent load they put on the ring is taken into account in assessing the number of socket outlets that can be permitted, but these items have a different function from the general purpose socket outlets and it is logical to serve them separately. Separation by function can be an asset to maintenance; there should be no need to lay dead all the socket outlets in part of a building when work has to be done to a toilet extract fan. On the other hand, a separate circuit to one small piece of equipment may seem an extravagance. No general rule can be made, and the designer must decide each application on its particular circumstances.

Equipment larger than 3 or 4 kW must in any case have a circuit for every individual item. This applies to cookers, each of which must be connected through a cooker control unit, and each of which must be on a circuit of its own. A cooker with two hot plates, a grill and an oven can take 35 A when everything in it is switched on, and in many households the cooker will be fully used, with all plates and oven on, at least once a week. In restaurant and school kitchens the cookers are likely to be in full use for the greater part of the time. Cooker control units are generally rated at 45 A and if they and the cookers are to be properly protected the circuit fuse must not be greater than 45 A. It follows that the circuit cannot serve anything in addition to the cooker control unit without being overloaded.

Other large equipment is likely to consist of motors driving pumps and fans in plant rooms and machine tools in workshops and factories. In the case of plant rooms each machine is almost invariably on a circuit of its own. Having more than one motor on a circuit would make it necessary to use very heavy cable and would in general be less economic than using a larger amount of smaller cable. It would also be extremely inconvenient to have several machines put out of action if one of them blows its fuse. This is particularly the case when one machine is intended as a standby for another. Similarly in factories it is usual to have each machine on a circuit of its own. In small and medium sized factories the most convenient wiring method is probably one using conduit and trunking. In such places there is seldom any objection to installing conduit and trunking on the surface of walls, and this is cheaper than burying it in the fabric of the building. It also makes it quite easy to alter the wiring when new machines are installed or the factory is rearranged. For the same reason, it is also better to run the wiring at high level under the ceiling and drop to the machines than to run it within the floor.

In large factories, a busbar system is often used. Bare conductors enclosed in a casing are run round the factory, preferably at high level, either on the walls or under the ceiling. A switch fuse is connected to these conductors as close as possible to each machine, and the connection from the switch fuse is taken through conduit or trunking to the machine. Each machine is thus on its own circuit, but no sub-mains other than the busbars are needed. The busbars must be protected by an adequate switch fuse at the intake. It is easy to connect a new switch fuse at any point of the busbars and the electrical installation is thus both convenient and flexible.

In small workshops, for example, metalwork and engineering rooms in secondary schools, the machines used may be small enough to make it practicable to serve a number of them from one ring circuit. Each machine is connected to the ring through a fused isolator or through a switch fuse. The fuse is necessary to protect the final connection to the machine, which is necessarily of a lower rating than the ring main, and to protect the internal wiring of the machine which will also be of smaller wire than the ring main. The cables of the ring main should be capable of carrying at least 70% of the total current taken by all the machines, and it will be found that this very soon restricts the size of workshop which can be treated in this way.

It should be appreciated that everything that has been said about power circuits applies equally to three phase and single phase circuits. Where three phase machines are used three or four wires, according to the system, plus an earth connection, are installed, and distribution boards, isolators and circuit breakers are of the three phase pattern, but the general circuit arrangements are the same as for single phase circuits.

All mechanical equipment requires maintenance, and all machines and equipment must, therefore, be installed in such a way that maintenance is possible. One of the things that has to be done before maintenance work is started is the turning off of the electricity supply, and it must be possible to isolate each machine or group of machines. It has been known to happen that an electrician has turned off an isolator in a switch room and gone to work on a machine some way from that room, that someone else has come along later, not realized that anyone was working on the machine and has turned the isolator on again. Not only has this happened, it has caused deaths. Consequently, most safety regulations now require that there should be an isolator within reach of the machine in addition to any means of isolation further away. The intention is that no-one can attempt to turn the supply on without the man on the machine becoming aware of what is happening. For small machines, such as roof extract fans, connecting the machine to the wiring through a socket and plug near the machine is a convenient and satisfactory way of providing local isolation. For larger machines, a switch or isolator has to be installed.

9 Protection

Introduction

It is a truism that electricity is dangerous and can cause accidents. A large part of any system design is concerned with ensuring that accidents will not happen, or that if they do, their effects will be limited. It might be reasonably argued that these considerations are the most important part of a design engineer's task. In the previous chapters, we have spoken about choice of accessories, selection of cables and their correct sizing, the arrangement of outlets on a number of separate circuits and the proper ways of installing cables and we have pointed out the need for protecting cables against mechanical damage. If these matters are given the care they deserve, the likelihood of faults on the electrical installation will be small. Nevertheless, it is still necessary to provide protection against such faults as may happen.

The general principle of protection is that a faulty circuit should be cut off from the supply and isolated until the fault can be found and repaired. The protective device must detect that there is a fault and must then isolate the part of the installation in which it has detected the fault. One could perhaps suggest many theoretical ways of doing this, but it is also necessary that the method adopted should bear a reasonable proportion to the cost of the whole installation. Historically, the methods which could be adopted at any time depended on what devices could be economically manufactured at that time, but once a method has been adopted it tends to remain in use and newer products do not completely supersede it. Enthusiasts sometimes stress the advantages of a new idea while forgetting that the older method had some favourable features which the new one does not match. The result of developments is that at present there are several protective devices available and there appear to be no overriding grounds for preferring any one to the others.

The devices available restrict the type of protection that can be given. A logically ideal system of protection against all possible faults cannot be made economically, and the protection designed must make use of the equipment commercially available. This can lead people to argue from the available techniques to the faults to be guarded against, and in the process become so obsessed with the ease of guarding against an improbable fault that they forget the importance of protection against a more likely one. It seems more satisfactory to start by considering the faults that may happen.

The two dangers to be prevented are fire and shock to people. In turn these dangers can arise from three kinds of fault, namely a short circuit, an overload and a fault to earth.

If through a fault in the wiring or in an appliance the line and neutral conductors become connected, the current that flows is limited only by the sum of the resistance of the cables of the permanent wiring and the impedance of the accidental contact between the two cables. If the fault connecting the line and neutral has a negligible impedance the two conductors are effectively short circuited. The current that flows through the conductors is very high and if allowed to continue would burn the insulation. The high conductor temperature resulting from the excessive current could start a fire. If the excess current continues to flow further after the insulation has been damaged, there is also a possibility that the conductor might touch exposed metal and give a shock to anyone touching the metal.

If the fault that connects the line and neutral has some impedance the current flowing through the fault and conductors is less than the current in a complete short circuit. It is still likely to be higher than the maximum current the circuit can safely carry and if it persists over a period of time, it can cause serious damage. When the fault current is much less than that of a complete short circuit, it is described as an overload. An overload can also be caused by an electric motor.

When an excessive mechanical load is imposed on an electric motor it continues to run but draws a higher than normal current from the supply. The circuit supplying the motor, therefore, carries a higher current than it has been designed for, and although it is not as high as a short circuit current, it can still be high enough to be dangerous. A fault in the internal wiring of a motor can also cause an electrical overload, although if it is serious enough it is likely to amount to a short circuit.

A fault to earth occurs if through some defect the line conductor becomes connected to earthed metalwork. The effect is similar to a short circuit, but whereas a short circuit will raise exposed metalwork above zero potential only after something further has happened after the short circuit occurs, increasing the potential of external metal is an immediate part of an earth fault itself. We can see this by looking at Fig. 82 which shows diagrammatically an electric fire with an earthed metal case. Suppose the fire becomes damaged and the live wire touches the case at point A. A current will flow through the case and earth wire

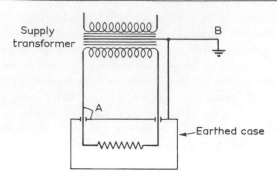

Fig. 82   Earth fault

to earth at point B, which would normally be the earth at the Electricity Board's transformer.

Now let E = supply voltage

I = fault current flowing

Z_T = total impedance from line connection of supply transformer through line conductor, fault and earth conductor to earth connection at supply transformer.

Z_E = impedance of earth path from fault back to earth connection.

Then the current flowing will be E/Z_T, and the voltage drop between A and B will be $IZ_E = E.Z_E/Z_T$. Now Z_E/Z_T is likely to be of the order of 0.4 to 0.5, so that on a 240 V supply the metal case at A will be raised to about 100 V.

We cannot explain how electrical circuits in buildings are protected against short circuits, overloads and earth faults without referring to the various protective devices which can be used. To make our account intelligible we propose first of all to describe the devices available and then go on to discuss how they are applied in practice.

Rewirable fuses

The earliest protective device consisted of a thin fuse wire held between terminals in a porcelain or bakelite holder. It is illustrated in Fig. 83. It is inserted in the circuit being protected and the size of fuse wire is matched to the rating of the circuit. The fuse is designed so that if the current exceeds the rated current of the circuit the fuse wire melts and interrupts the circuit.

Although commonly called rewirable fuses, their correct name is semi-enclosed fuses, and it is by this name that they are referred to in British Standards and in the I.E.E. Regulations.

H.R.C. fuses

The rewirable fuse has limited breaking capacity. If a very large current flows the

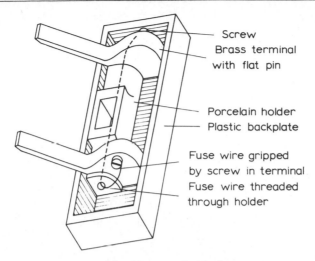

Screw
Brass terminal
with flat pin

Porcelain holder
Plastic backplate

Fuse wire gripped
by screw in terminal
Fuse wire threaded
through holder

Fig. 83   Rewirable fuse

fuse wire melts very rapidly and a large amount of energy is released. It can be large enough to cause serious damage to the fuse carrier. It was found that some of this energy can be absorbed by a packing of inert fibrous or granular material wound with wire, and this led to the development of the cartridge fuse, illustrated in Fig. 84.

The fuse wire is mounted between two end caps which form the terminals of the complete fuse link. The wire is surrounded by a closely packed granular filler and the whole is contained in a solid casing. When the wire melts, or blows, the energy is absorbed by the granular filler. Fuses of this type are known variously as High Rupturing Capacity (H.R.C.) or High Breaking Capacity (H.B.C.) Fuses, or less technically as cartridge fuses.

Operation of fuses

Both rewirable and cartridge fuses work in the same way. The current heats the fuse wire until the latter melts, after which there is an arc between the ends of the broken wire, and finally the arc extinguishes and the circuit is completely interrupted.

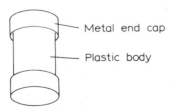

Metal end cap

Plastic body

Fig. 84   Cartridge fuse

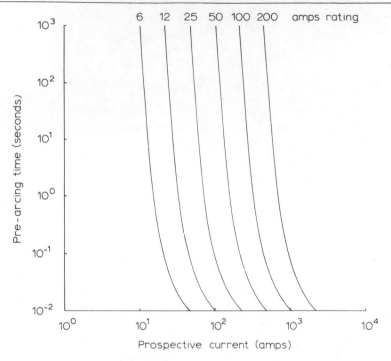

Fig. 85   Time current characteristics of HRC fuses

The time taken for the fuse to melt depends on the magnitude of the current, and a fuse will have a characteristic curve of time against current. A set of such characteristic curves is shown in Fig. 85. The total operating time is the sum of the melting, or pre-arcing, time and the time during which there is an arc, known as the arcing time. The arcing time varies with the power factor and transient characteristics of the circuit, the voltage, the point in the alternating cycle of supply at which the arcing commences and on some other factors. It is not, however, of significant length except for very large overcurrents when the total operating time is very short.

The *minimum fusing current* is the minimum current at which a fuse will melt, that is to say the asymptotic value of the current shown on the time-current characteristics. The *current rating* is the normal current. It is the current stated by the manufacturer as the current which the fuse will carry continuously without deterioration. It is also referred to as current carrying capacity and other similar terms. The *fusing factor* is the ratio

$$\frac{\text{minimum fusing current}}{\text{current rating}}$$

When a short circuit occurs, the melting process is adiabatic and the melting energy is given by

$$W = \int_0^{t_m} i^2 R \, dt$$

where W = melting energy

i = instantaneous current

R = instantaneous resistance of that part of element which melts on short circuit.

t = time

t_m = melting time

R is assumed to vary in the same manner with i and t for all short circuits and the quantity

$$\int_0^{t_m} i^2 \, dt$$

is approximately constant for the pre-arcing time of a fuse. It is often called the pre-arcing I^2t. It is this quantity which determines the amount of excess energy

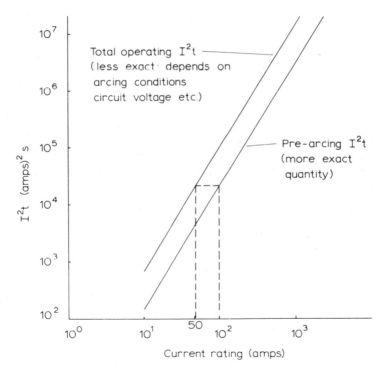

Fig. 86   Short circuit I^2t characteristics

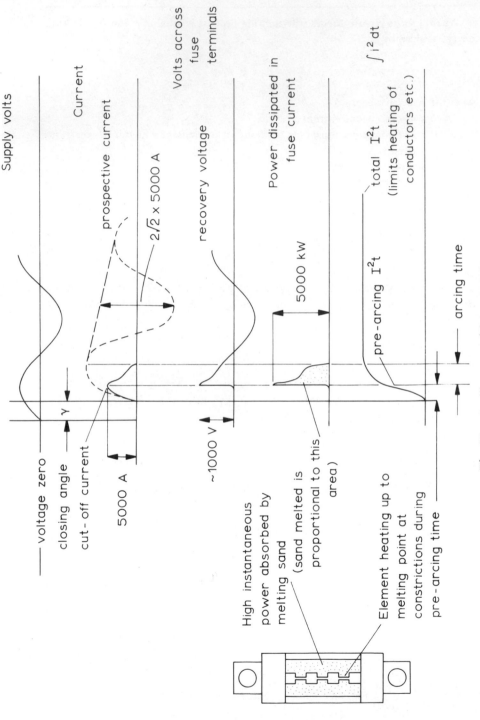

Fig. 87 Oscillograms of fuse operation

Supply volts

Current

prospective current

$2\sqrt{2} \times 5000$ A

Volts across fuse terminals

recovery voltage

Power dissipated in fuse current

$\int i^2 dt$

total I^2t
(limits heating of conductors etc.)

pre-arcing I^2t

5000 kw

arcing time

voltage zero

closing angle

cut-off current

γ

5000 A

~1000 V

High instantaneous power absorbed by melting sand (sand melted is proportional to this area)

Element heating up to melting point at constrictions during pre-arcing time passing through the circuit before the circuit is broken and it is particularly important in the protection of semiconductor circuits and the reduction of overheating in power circuits. Typical $I^2 t$ characteristics are shown in Fig. 86.

Oscillograms of the operation of a fuse are shown in Fig. 87.

M.C.Bs

An alternative to a wire which melts when overheated is a circuit breaker. A miniature circuit breaker is one which has a rating similar to that of a fuse and is about the same physical size as a fuse carrier of the same rating. A typical miniature circuit breaker (M.C.B.) is shown in Fig. 88.

It has a magnetic hydraulic time delay, and the essential component is a sealed tube filled with silicone fluid which contains a closely fitting iron slug. Under normal operating conditions the time delay spring keeps the slug at one end of the tube (a).

When an overload occurs, the magnetic pull of the coil surrounding the tube increases, and the slug moves through the tube, the speed of travel depending on

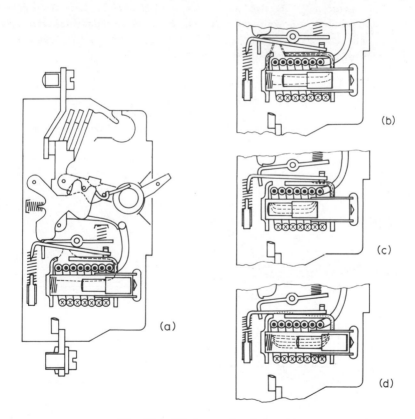

Fig. 88   Miniature circuit breaker

the magnetic force and, therefore, on the size of the current, (b). As the slug approaches the other end of the tube the air gaps in the magnetic circuit are reduced and the magnetic force is increased until it is great enough to trip the circuit breaker, (c). With this mechanism the time taken to trip is inversely proportional to the magnitude of the overload.

When a heavy overload or a complete short circuit occurs, the magnetic force is sufficient to trip the circuit breaker instantaneously in spite of the large air gaps in the magnetic circuit, (d). In this way, time delay tripping is achieved up to about seven times rated current, and instantaneous tripping above that level.

An alternative design has a bi-metal element which is heated by the circuit current and operates the trip catch when it deflects. A simple magnetic coil is included to trip the catch on higher overloads, so that the resulting characteristic is similar to that of the magnetic-hydraulic type. The thermal-magnetic type is liable to be affected by the ambient temperature. If several M.C.Bs are mounted inside a closed distribution board, the heat from the currents passing through all the circuits in the board will raise the temperature inside the enclosure. Thermally operated M.C.Bs used in this way may have to be de-rated to prevent their tripping before an overload occurs. The effect of ambient temperature can, however, be reduced and the need for de-rating obviated by designing the bi-metal to run at a relatively high temperature.

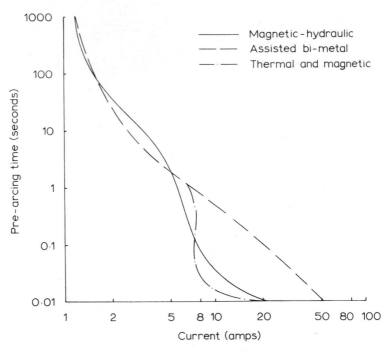

Fig. 89   Time-current curves for M.C.Bs

Typical time-current characteristics of M.C.Bs are given in Fig. 89.

The M.C.B. has a toggle switch by which it can be operated manually. This switch is thrown into the off position when the overload device trips the breaker, and the M.C.B. is reset by the same switch. M.C.Bs can, therefore, combine the functions of switch and fuse, and in some cases this is a very useful and economic procedure. In a factory or store, for example, one may want to control the lights for a large area from a bank of switches at a single point. If a distribution board with M.C.Bs. is placed at this point, it is possible to dispense with a separate bank of switches.

E.L.C.B.

Another device frequently used is the earth leakage circuit breaker (E.L.C.B.). This is a circuit breaker which detects a current leaking to earth and uses this leakage current to operate the tripping mechanism. The leakage current is a residual current and a more general name for the device, which is used in the I.E.E. Regulations, is residual current device. There are two types, namely current operated and voltage operated.

The principle of the current operated E.L.C.B. is shown in Fig. 90. The load current through the circuit is fed through two equal and opposing coils wound on a common transformer core. On a healthy circuit, the line and neutral currents are the same and produce equal and opposing fluxes in the transformer core. However, if there is an earth fault, more current flows in the line than returns in the neutral, and the line coil produces a bigger flux than the neutral

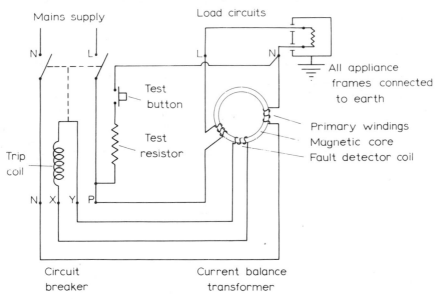

Fig. 90　Current operated earth leakage circuit breaker

coil. There is thus a resultant flux which induces a current in the search coil, and this in turn operates the relay and trips the breaker.

The value normally adopted for the rated tripping current, which is defined as the out of balance current at which the circuit breaker will trip in less than 0.1 s, is 500 mA. The unit always includes a test switch which simulates an out of balance condition by injecting a test current bypassing one of the principal coils. A resistor limits the magnitude of the test current so that the test also checks the sensitivity of the breaker. It should be noted that the test switch checks the operation of the earth leakage circuit breaker, but does not check the soundness of the earth continuity conductor.

The principle of the voltage operated E.L.C.B. is shown in Fig. 91. It will be seen that the trip coil operates if a sufficient voltage appears on the earthed metalwork. Standard units are designed to operate before the voltage at the earth point reaches 40 V. A test switch is again provided. The primary function of a voltage operated earth leakage circuit breaker is to give protection against shock, and this is done by the detection of any dangerous voltage rise on the metalwork of the system. In effect, the circuit breaker is a voltmeter measuring the potential difference between the protected metalwork and earth. The earth end of the trip coil must, therefore, be connected to an earth electrode. To be effective such an electrode must be outside the resistance area of any other electrode. The provision and connection of a good earth electrode are part of the cost of installing a voltage operated E.L.C.B.; we shall describe the construction of earth electrodes later in this chapter.

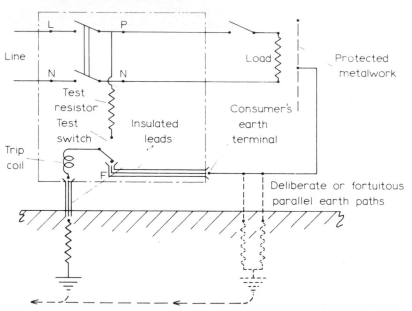

Fig. 91   Voltage operated earth leakage circuit breaker

Isolating transformers

In some applications a double wound transformer can achieve adequate protection for the system on its secondary side. The secondary is not earthed and the construction of the transformer is carefully designed to prevent any possibility of contact between the secondary and primary windings. The voltage between the line and return connections of the secondary is fixed, but because no part of it is earthed or connected to any other fixed point, the voltage relative to earth is quite undetermined. In other words the line can be at 240 V above earth and the return at earth, or the one can be at 120 V above earth and the other at 120 V below earth, or the line can be at earth and the return at 240 V below earth, or any other combination.

Thus, if anyone in contact with earthed metal touches the line conductor, he immediately brings the line to earth potential and the return drops to 240 V below earth. The same thing happens if through a fault in an appliance a conductor touches earthed metal. This prevents danger of shock. It is important to note that the system will not ensure safety if more than one appliance is fed from the transformer. Fig. 92 shows two appliances near some earthed metalwork. Suppose a fault develops at A. The line conductor will be held at earth potential and the return conductor will be at mains voltage below earth. The circuit continues in operation and there is now no protection against a second fault occurring at another point such as B.

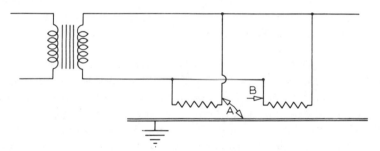

Fig. 92   Wrong use of isolating transformer

The chief use of isolating transformers is in shaver units fixed in bathrooms to supply electric razors. There is practically no possibility of two independent faults occurring in such an application.

We have now reviewed the equipment which is available to protect electrical installations in buildings and can go on to consider the nature of the protection that is needed.

Capacity of circuit

The methods by which a cable is sized for the duty it has to perform have been explained in Chapter 4. There is no need to be worried about the effects of a fault on the voltage drop; it is the high current that has to be guarded against. The protection has to safeguard the circuit against excessive currents, and an excessive current is any current higher than that for which the circuit has been designed. Now the cable must be rated at a little more than the current actually taken by the circuit, and it is the current carrying capacity of the cable that determines the protection which has to be provided. In other words, the current rating of the fuse or circuit breaker protecting the circuit must be the normal current which the circuit cable has been designed to carry.

The guiding principle to be followed is that every cable in a permanent installation in a building must be protected. The protective device must not have a current rating greater than that of the cable, and in most cases it will have a rating either equal to or only just less than that of the cable. It follows that at every point at which a smaller cable branches from a larger one there must be a protective device to safeguard the smaller cable. In conventional systems, this is provided by the use of switchgear and distribution boards where a main divides into two or more sub-mains and where a sub-main divides into a number of final circuits. It is also because every branch has to be protected that a fused spur unit is installed whenever a single branch is taken off a ring main.

Fault currents

The current normally plotted on the time-current characteristics of fuses and circuit breakers is that known as the *prospective fault current*. This is the current which would flow in the circuit if the fuse were not there and a complete short circuit occurred. It is indicated as a dotted line in Fig. 87. In practice, the fuse will open the circuit before this prospective current is reached; the fuse is said to cut off and the instantaneous current attained is called the cut off current.

The wave form of the prospective fault current depends on the position of the fault within the whole of the supply network, the relative loading of the phases within the network and whether the supply comes through a transformer or directly from a local generator. These questions can be of importance to the engineer concerned with protecting a public supply system, but are of less consequence to the designer of the services within a building. For his purposes the simple procedure described here is adequate and more complicated considerations need not be taken into account in selecting the protection devices to be used within the building.

The prospective current is defined as the R.M.S. value of the alternating component, whereas the cut off current is defined as the instantaneous current at cut off. These definitions produce the paradox that the numerical value of the cut off current may be greater than the numerical value of the prospective current.

The prospective fault current is determined by the supply voltage and the impedance of the path taken by the fault current. In Fig. 93, this path is indicated by a b c d e f g h. The figure shows diagramatically the usual situation in urban installations where the consumer's earth point is connected to the sheath of the Electricity Board's cable, which is earthed at the transformer end. If the supply voltage is E and the impedance of this path is Z_f, then the prospective fault current is given by $I_f = E/Z_f$. The total impedance Z_f is made up of the impedance of the transformer and the impedance of all the cables in the path. The impedance of a transformer is almost entirely reactive, and is therefore always referred to as reactance. In practice, it is found that the reactance of the transformer sets an upper limit to the fault current that can flow. Thus if a normal 500 kVA transformer is short circuited the current on the secondary side will be 14 000 A, while the corresponding figure for a 750 kVA transformer is 21 000 A. These are average figures for typical transformers and ignore the impedance of the supply system on the primary side of the transformer. They therefore include a small hidden safety factor. The resistance of the service cable and its sheath is usually quite low, but even a short length of sub-circuit cable makes a great reduction in the prospective fault current. For example 20 m of 1.5 mm^2 cable in an installation supplied from a 750 kVA transformer will limit the fault current to 1000 A.

Data on cable resistance are given by the manufacturers in their catalogues, and the leading makers of fuses and M.C.Bs provide in their catalogues tables and graphs showing the prospective fault currents with different lengths of cable on the secondary side of standard transformers. The fault itself usually has some impedance, so that the actual fault current is less than the prospective fault current. The prospective fault current is calculated on the assumption that the impedance between d and e in Fig. 93 is zero. The actual fault current could theoretically be calculated by adding the actual impedance of the fault between d and e to the impedance used for calculating the prospective current, and as the total actual impedance will thus be higher than that of the complete short circuit assumed in calculating the prospective current, the actual fault current will be less than the prospective one.

When a fault occurs on an appliance the cables of the final sub-circuit play an important part in limiting the fault current. Nevertheless, there is always the

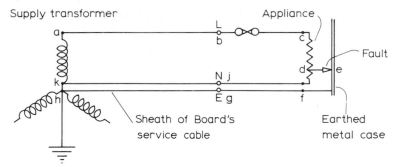

Fig. 93   Fault currents

possibility, even if a small one, of a low impedance fault immediately behind the fuse or circuit breaker. It is this fault which, however unlikely it may be that it will happen, will produce the highest possible fault current and it is this fault which the protection ought to be capable of dealing with.

Discrimination

In an installation having the type of distribution described in Chapter 6, there is a series of fuses and circuit breakers between the incoming supply and the final outlet. Ideally, the protective devices should be graded so that when a fault occurs only the device nearest the fault operates. The others should not react and should remain in circuit to go on supplying other healthy circuits. Discrimination is said to take place when the smaller fuse opens before the larger fuse operates. This is the desired state of affairs, and when the unwanted converse situation happens it is said that discrimination is lost.

Fig. 94 shows the time current characteristic for an M.C.B. serving a final sub-circuit and for a 60 A H.B.C. fuse protecting the sub-main leading to the distribution board on which the M.C.B. is mounted. If the fault current is less than 1200 A, the M.C.B. will open first. If the fault current is greater than this, the H.B.C. fuse will blow before the M.C.B. can operate. The resistance of the cables of the sub-circuit reduces the fault current, so that the further away from the distribution board a fault occurs, the lower will be the fault current. If the fault current at the distribution board is, say, 2000 A, it will fall below this along the sub-circuit from the board and will drop to 1200 A fairly soon after the board. A fault near the board would blow the board fuse but a fault on the sub-circuit any distance from the board would leave the board intact and open the M.C.B. The discrimination is said to be acceptable. If the fault current at the end of the sub-circuit is so high that the H.B.C. fuse always opens before the M.C.B., there is no discrimination. If there is discrimination at the end of the sub-circuit but not for a fault a short way before the end of the sub-circuit, we say the discrimination is not acceptable. With the characteristic curves of Fig. 94, the discrimination would probably be acceptable even for a fault level at the board of 3000 A. The possibility of a fault at the board is small and the actual fault currents to be cleared will almost certainly be due to faults at appliances. Provided discrimination is maintained for these faults, its loss on the rare occasions when a fault occurs on the permanent wiring close to the board can probably be accepted.

In general, discrimination is a problem only when a system uses a mixture of M.C.Bs and fuses. It can be seen from Figs. 85 and 86 that a fuse will always discriminate against another fuse of a larger rating.

Breaking capacity and back up protection

A certain amount of energy is released when a circuit is broken, whether by a fuse or by a circuit breaker. This energy must be absorbed by the device

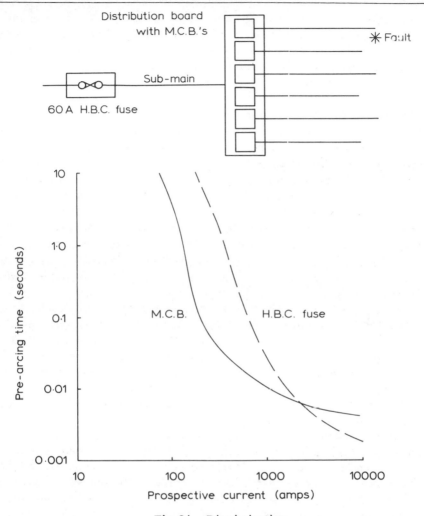

Fig. 94   Discrimination

breaking the circuit, and the capacity to do so limits the current which the device can safely break. The breaking capacity is defined as the maximum current that can be broken at a stated voltage and power factor. In switchgear practice, it is common to refer to breaking capacity in MVA, but for the fuses and circuit breakers used in building services, it is usual to refer to breaking capacity in amperes. However, whether MVA or amperes are used, the statement of breaking capacity is incomplete unless it contains the voltage and power factor at which it applies.

In building services, in practice faults are invariably so close to unity power factor that it is hardly necessary to specify power factor. Fuses and circuit breakers are rated at the voltage of the supply which in the United Kingdom is either 240 V or 415 V, and therefore the statement of voltage can be taken for granted. Thus, in spite of what has been said in the last paragraph breaking capacity is often quoted in amperes only, without further statement.

The cut off current depends not only on the characteristics of the fuse but also on the nature of the fault current and the point in the alternating supply cycle at which the fault occurs and the fuse starts to act. Although in the majority of cases the maximum prospective fault current will not in fact be reached, for safety the fuse or circuit breaker must have a breaking capacity at least equal to the maximum prospective fault current. The breaking capacity of the various devices used in protection is of some importance and must be taken into account when a selection has to be made between different devices and schemes of protection.

It is not easy to give an exact figure for the breaking capacity of a rewirable fuse, because it depends on the type of fuse carrier or fuse holder used and the exact composition of the fuse wire and also because it is liable to vary with the age of the fuse wire. As an approximation it can be taken to be of the order of 3000 to 4000 A.

H.R.C. fuses to B.S. 1361:1971 have a breaking capacity of 16 500 A at 0.3 power factor when rated at 240 V and of 33 000 A at 0.3 power factor when rated at 415 V. M.C.Bs manufactured in the United Kingdom in the ratings used on distribution boards to protect final sub-circuits generally have a breaking capacity of 3000 A, although some German makes for similar applications have far higher capacities.

It may often be convenient to use a breaking device which has breaking capacity less than the maximum prospective fault current.

As an example, with the arrangement of Fig. 94, it could be that the fault current for a dead short circuit on an appliance at the end of a sub-circuit is, say, 1500 A, whilst the fault current for a short circuit on the wiring within a short distance of the board is 4000 A. The former is the only fault which is probable, and an M.C.B. with a breaking capacity of 3000 A might well be considered a suitable form of protection. If this were installed and the possible fault of 4000 A occurred and the M.C.B. were left to clear it, the M.C.B. would suffer damage and the wiring might also be damaged before the circuit was fully broken. In such a case *back up protection* is required, and is provided by the H.B.C. fuse at the other end of the sub-main. The fuse limits the maximum fault energy; if the fault current and, therefore, the fault energy is greater than the M.C.B. can handle the back up fuse blows. If on the other hand the fault current is within the capacity of the M.C.B. the back up fuse does not act.

Back up protection and discrimination are closely connected, but they are not the same thing. Even if the breaking capacity of the final sub-circuit fuse is such that no back up protection is needed, the sub-circuit fuse must still discriminate against the sub-main fuse. In other words, discrimination is still needed even when back up protection is not. Since discrimination is always needed it must also accompany back up protection, but it does not follow that the provision of back up protection will automatically give discrimination. It is quite possible to choose the back up fuse so that it blows before the sub-circuit fuse, however small the fault current, and this would be back up protection without discrimination. It is of course to be avoided.

Summing up, we can say that fuses and M.C.Bs must be so chosen that back up protection is provided if it is needed and that discrimination is always provided.

We have described the equipment that is in practice available to give protection to electrical installations and we have considered the nature and magnitude of fault currents and the interaction of fuses and circuit breakers in series with each other. Knowing the probable faults and the equipment available we can now turn to consider how this equipment can be used to give satisfactory protection against the faults that may arise.

Overload protection

Fuses and M.C.Bs react to short circuits and thus provide protection against their consequences. However, their time-current characteristics are such that they do not provide real protection against sustained low level overloads. For example, a 30 A H.B.C. fuse will carry 40 A indefinitely. This will not harm the cables of the permanent wiring if they have been correctly sized and the fuse has been correctly matched to them An electric motor overloaded to this extent would however burn out.

The I.E.E. Regulations require overload protection which will operate before the current exceeds 1.45 times the current-carrying capacity of the smallest conductor in the circuit. If protection against smaller overloads than this is needed it has to be supplied with the equipment which needs the protection. The most usual case is that of an electric motor which is protected against overload by an overload relay in the motor starter. The usual type of overload relay contains a heater element and a bi-metal strip in each motor line conductor. An excessive current causes the bi-metallic strip to deflect, the amount of deflection depending on the magnitude of the current and the time for which it flows. When the deflection reaches a predetermined amount the bi-metal operates the tripping mechanism which opens the coil circuit and this in turn causes the main contacts to open.

Such an overload protection has a time-current characteristic of the form shown in Fig. 95. It will be seen that the starter with this overload device will carry a starting current 8 times the full load current for 6s. A fuse of the same rating would not do this.

A motor starter must be capable of operating several thousand times in its working life. A design which enables it do so places a limit on the magnitude of the current it can break. Thus while it can deal with overloads it cannot safely break a short circuit. For this reason, the starter which protects the motor

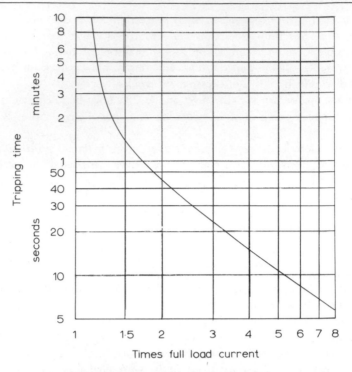

Fig. 95   Motor starter characteristics

against overloads must itself be protected by a fuse to deal with short circuits. The fuse will at the same time protect the permanent wiring against short circuits.

Now a fuse of the same rating as the motor will not carry the starting current for long enough for the motor to run up to speed. The fuse backing up the starter must, therefore, have a rating higher than the normal full load running current of the motor. Motor manufacturers and starter manufacturers provide information which enables the fuse to be correctly chosen. Table 1 is a selection from such data.

The arrangement of fuse, starter and motor is shown in single line diagrammatic form in Fig. 96. Logically since cables ab and cd are protected by the fuse, they should not have a current rating less than that of the fuse, but this rating is much more than the normal running current and cables so chosen will

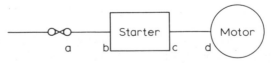

Fig. 96   Motor protection

Table 1 Fuse ratings for motor circuits

Type of starter	Overload release rating-amps	Fuse rating amps
Direct on	0.6 to 1.2	5
	1.0 to 2.0	10
	1.5 to 3.0	10
	2.0 to 4.0	15
	3.0 to 6.0	20
	5.0 to 10.0	30
	9.0 to 15.0	40
	13.0 to 17.0	50
Star-delta	4.0 to 7.0	15
	6.0 to 10.0	20
	9.0 to 17.0	30
	16.0 to 26.0	40
	22.0 to 28.0	50

be unnecessarily heavy. For example, we can see from Table 1 that a motor with a running current of 3 A and a starter overload set at 3 A will need a back up fuse of 15 A. To use cable rated at 15 A for a circuit which carried a normal current of 3 A would be uneconomic.

One can, however, take advantage of the fact that the starter protects the cable as well as the motor. In theory, if a very high resistance fault developed on the cable between a and b neither the fuse nor the starter would open the circuit, and damage might result. In practice, the chances of this happening are so small that they may be neglected. The usual practice is to make cables ab and cd such that their current rating is half that of the fuse. This is probably the only case in which it is in order for the cable to have a rating less than that of the fuse protecting it, and it may be done only where there is a motor starter incorporating an overload relay.

Protection of persons

People using the building have to be protected against electric shock. They would get a shock if they came into contact with live parts, and in considering protection a distinction is made between direct and indirect contact. Direct contact is contact with a live conductor which is intended to carry current. The normal protection against this is the provision of insulation on all current-carrying wires.

Indirect contact is contact with exposed metal parts which are not intended to carry current but have become live as a result of a fault. Such a fault is indicated in Fig. 93. When it occurs, the metal case of an appliance which a person is likely to handle is raised to line potential and will cause a shock if it is touched by someone using the appliance. Protection is provided by the fact that the case is earthed and that a protective device will disconnect the circuit as soon as a fault current flows to earth.

It is possible that a fault of this type will occur while a person is holding an appliance. In that case he will be subjected to a dangerous potential during the time it takes for the protective device to operate and disconnect the supply. For this reason the I.E.E. Regulations stipulate the time in which the device must operate, which is 0.4 seconds for circuits serving socket outlets and 5 seconds for circuits serving only fixed equipment. The reason for the difference is that a person is likely to have a firmer grip on a portable appliance plugged into a socket outlet than on a fixed appliance which he is only likely to touch casually.

A current would also flow through a person if he came into contact with two separate pieces of metal at different potentials. In the event of a fault such a potential difference could exist between the exposed metal case of an electrical appliance and metal which is not part of the electrical system, such as a sink or central heating radiator. To avoid danger arising in this way all exposed metal must be linked to the protective earthing point of the installation. The conductor by which this is done is known as the equipotential bonding conductor and methods for determining its size are given in the I.E.E. Regulations.

Earth protection
As we have seen, fuses and M.C.B's react to short circuits. If they are to provide protection against faults to exposed metalwork, the wiring must be such that the fault produces the same conditions as a short circuit, namely a large excess current in the line conductor. This will happen only if the impedance of the path taken by the fault current is low enough. This path is indicated by a b c d e f g h in Fig. 93 and is known as the earth fault loop; its impedance is known as the earth fault loop impedance.

An earth fault occurs when a live conductor touches exposed metalwork and so raises the metal to a dangerous potential. If a fuse or M.C.B. is to be used to clear a fault of this nature, then the fault must immediately produce a current large enough to operate the fuse or breaker. To achieve this, all exposed metalwork which, in the event of some fault, could conceivably become live, is earthed and the earth path is designed to have a low impedance. When a fault occurs, current flows through it to earth, and because of the low impedance in the earth fault loop path, the current is large enough to operate the protective device. It must also be large enough to operate the device within the time stipulated in the I.E.E. Regulations. The Regulations give details of the maximum earth fault loop impedance which can be allowed to ensure operation within the required times for fuses, M.C.B's and E.L.C.B's.

Since an earth fault raises exposed metal above earth potential it creates the possibility of sparking from such metal to nearby metal not affected by the fault and therefore still at earth potential. Although the fuse or M.C.B. will clear the fault, that is to say remove it, the operating time of the fuse or M.C.B. may be long enough to allow a spark to occur at the fault position. The only sure way to prevent sparking over from exposed metalwork to adjacent metal of other services is to bond the two sets of metalwork together. We have already mentioned this in Chapter 6. In any event the Institution of Electrical Engineers' Regulations require the main earth terminal at the incoming supply intake to be bonded ;o the metalwork of any gas or water services as near as possible to the point at which those services enter the building.

The policy of protecting against earth faults by making sure that they produce a short circuit and, therefore, operate the fuse or circuit breaker has been described as 'chasing the ampere'. With ever increasing demand for electricity, circuit ratings and their fuse ratings are becoming ever higher and the short circuit current needed to operate the fuse or circuit breaker becomes higher. Thus ever lower values are required for earth loop impedances and the fault currents which the fuses have to break become higher. The opinion has been expressed that this method of protection will not be able to keep pace with the consequences of increasing electrical loading.

However, at the present time in urban areas where the Electricity Board provides an effective earth to the sheath of the service cable, there is no difficulty in achieving an earth loop impedance of less than 1.0 ohm. It may perhaps not be possible to do this as long as steel conduit is relied on as the earth return path, but it can certainly be done by the use of a separate earth wire. Because of the need for low earth impedances separate earth wires should always be used on new installations.

If for any reason the earth loop impedance cannot be made low enough, it becomes necessary to use an earth leakage circuit breaker. Since an E.L.C.B. will operate on an earth fault current of 500 mA, it will react on a 240 V circuit even if the earth loop impedance is 480 ohms and it is inconceivable that values anything like as high as this would ever be found on a practical installation.

Temporary installations are likely to have higher earth loop impedance paths than permanent ones, and it is very common to use earth leakage protection for the temporary wiring on building sites, particularly in rural areas.

It should be noticed that earthing of exposed metalwork is still necessary when an E.L.C.B. is used. This can be shown by reference to Fig. 93. Suppose the earth return path f g h did not exist, then normal current would flow through a b c d j k and there would be no unbalance between line and neutral for a current operated E.L.C.B. to detect. Nevertheless the exposed metalwork, connected through the fault to the conductor at d, would be at a potential given by

$$\frac{\text{impedance of abcd}}{\text{impedance of abcdjk}} \times \text{line voltage}$$

and this could be anything up to almost full line voltage. A return path must be provided, but because of the sensitivity of the E.L.C.B. it need not be of particularly low impedance so long as it is continuous.

In the case of a voltage operated E.L.C.B. the metalwork must be connected to the trip coil of the circuit breaker, so that again an earth continuity conductor is needed.

Reliance on low earth loop impedance does not give protection against a fault which itself has a high impedance. Such a fault does not give rise to any danger of shock, because the impedance of the fault itself limits the fault current and, therefore, the potential to which the exposed metalwork can rise. We refer yet again to Fig. 93 and suppose this time that the impedance between d and e is Z_f whilst the impedance of the permanent wiring a b c d is Z_1 and that of e f g h is Z_r. The the fault current is:

$$I_f = \frac{E}{Z_f + Z_1 + Z_r}$$

where E = line voltage, and the potential drop from a to e is

$$P = I_f(Z_f + Z_1)$$

$$= \frac{E(Z_f + Z_1)}{Z_f + Z_1 + Z_r}$$

If Z_f is high, Z_r can be neglected.

$$\therefore \quad P = E$$

But the voltage at e is line voltage less the potential drop from a to e.

$$\therefore \quad \text{voltage at e} = E - P = 0$$

Nevertheless, such a fault, particularly on the peramanent wiring rather than on an appliance, could conceivably produce local overheating and cause a fire. Reliance on fuses and low earth loop impedances will not prevent this, but the use of a current operated earth leakage circuit breaker would. Although the probability of a fault of this type is very small, it is a reason for considering the use of E.L.C.Bs. The I.E.E. Regulations require an E.L.C.B. to be used to protect a socket outlet which is intended for portable equipment to be used outside the building. They also require all socket outlets to be protected by E.L.C.B's when earthing is provided by the user's own electrode and not by connection to an earth conductor provided by the Supply Authority. It is left to the designer's discretion whether a separate E.L.C.B. is provided for each socket outlet, whether each circuit is protected by an E.L.C.B. or whether all circuits on a distribution board are protected by an E.L.C.B. on the supply side of the board.

It is the designer's responsibility to use his own experience and professional judgement in selecting the scheme of protection.

Earth monitoring

Protection through earthing of exposed metalwork will fail if there is a break in the earth continuity conductor. It is possible to add an extra wire which will monitor the earth continuity conductor and break the main circuit if the earth continuity fails. The basic scheme of such a circuit is shown in Fig. 97.

The monitoring circuit incorporates a low voltage transformer, a relay, and a protective circuit breaker. An additional pilot lead is required to the appliance or portable tool. The low voltage is used to drive a current round the loop formed by the earth conductor, a section of the metallic housing of the appliance, the pilot conductor and the relay coil. This current holds in the relay and thus keeps the coil of the circuit breaker energized.

If there is a break in the earth continuity conductor, or indeed anywhere in the pilot circuit, the low voltage current fails, the relay is de-energized, the main contactor coil becomes de-energized and the circuit breaker opens the main circuit and cuts off the supply to the appliance. It will be noticed that this monitoring circuit merely checks that the earth continuity conductor is sound; it does not add to the basic earth leakage protection.

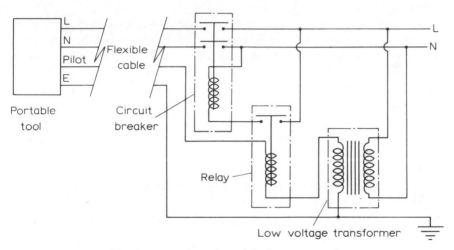

Fig. 97   Monitored earth leakage protection

Earth electrodes

In normal earthing, the earth and the neutral are quite separate. The load current flowing through the neutral must cause a potential difference between the two ends of the neutral. Since the end at the supply transformer is earthed, the end at the consumer's service terminal must inevitably be at some potential above earth. It cannot, therefore, be used as an earth point.

Nevertheless, an effective earth has to be found for the earth continuity conductors of the permanent installation in a building. In urban areas the sheath of the Electricity Board's service cable is normally used for this purpose, but there is no obligation on the Board to provide an earth, and in rural areas where the supply may be by overhead cable, it may not be possible for them to do so. In such cases, the consumer must provide his own earth electrode and the design of it becomes part of the design of the building installation.

An earth electrode is a metal rod, which makes effective contact with the general mass of earth. A common type consists of a small diameter copper rod which can be easily driven to a depth of 6 m or more into ground reasonably free of stones or rock. The soil remains practically undisturbed and in very close contact with the electrode surface. Since resistivity is lower in the deeper strata of earth and not very affected by seasonal conditions deep driving gives a good earth. Rods of this type are practically incorrodible. Also it is easy to get access to the connection at the top of the electrode. A typical arrangement is illustrated in Fig. 98.

Where the ground is shallow but has low resistivity near the surface, a plate electrode, either of copper or of cast iron, can be used. When the soil resistivity is high, a cast iron plate can be used with a coke surround. This method is illustrated in Fig. 99.

Standard cast iron plates are made for use as earth electrodes. They are complete with terminals for the earth continuity conductor. These terminals consist of two copper sockets each secured by a drift-pin, the two being joined by a tinned copper strand to which the earth conductor is bound and soldered. The completed connection is sealed and covered in bitumen before the electrode is buried. It is, therefore, not as accessible as the connection of the rod type electrode.

Long copper strip can also be used as an earth electrode, and the method of doing this is shown in Fig. 100. It will be seen from this that the strip is a useful type of electrode for shallow soil overlying rock. Strip may be arranged in single lengths, parallel lengths or in radial groups. Standard strip is commercially available for use as earth electrodes.

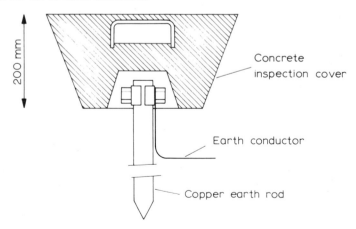

Fig. 98   Copper rod electrode

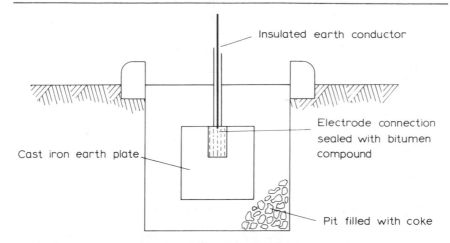

Fig. 99 Cast iron plate electrode

When current flows from the electrode into the soil, it has to overcome the resistance of the soil immediately adjacent to the electrode. The path of the current is shown in Fig. 101. The effect is equivalent to a resistance between the electrode and the general mass of earth, and this resistance is the resistance of the electrode. Furthermore, the surface of the ground near the electrode becomes live when current flows from the electrode to earth and Fig. 102 shows a typical surface distribution near a rod electrode. It can be seen that an animal standing near such an electrode could have a substantial voltage applied between its fore and hind legs, and in fact fatal accidents to livestock from this cause have been known. The earth electrode should, therefore, be positioned well out of harm's way. It should perhaps be noted that the deeper the electrode is below the ground, the smaller will be the voltage gradient at the surface.

The effectiveness of earth protection depends on the low resistance of the electrode when current flows through the electrode into the soil. This resistance cannot be accurately predicted in advance and must be checked by testing. After installation, the electrode should be periodically examined and tested to ensure

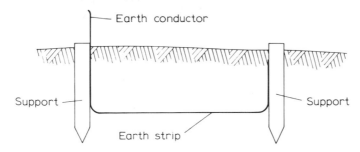

Fig. 100 Copper strip electrode

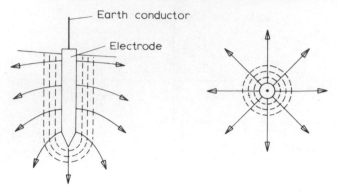

Fig. 101   Current from electrode into earth

that its initial low resistance is being maintained. The scheme for testing an electrode is shown in Fig. 103. The electrode under test is indicated by X; two auxiliary electrodes, Y and Z are driven in for the test. Y must be placed sufficiently far from X for the resistance areas not to overlap, and Z is placed approximately half way between X and Y. The test electrode X is disconnected from its normal continuity conductor and connected to the test instrument as

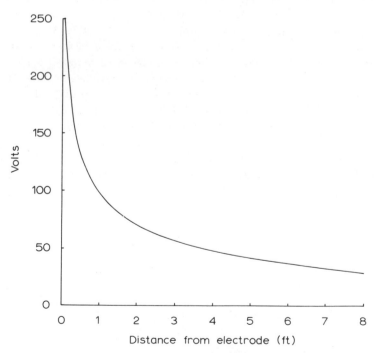

Fig. 102   Voltage at surface of ground due to rod electrode

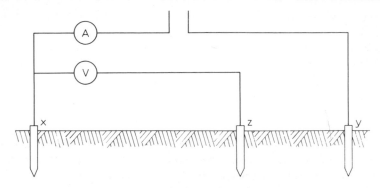

Fig. 103   Earth electrode resistance test

shown in Fig. 103. A low voltage alternating current is passed between X and Y. The current is measured and so is the potential between X and Z. The resistance of the earth electrode is given by the dividend of voltage and current. Check readings are taken with the electrode Z nearer to and further from electrode X, and the results are accepted only if all three readings are substantially the same. If they are not, the test must be repeated with a greater distance between X and Y.

Protective multiple earthing

This is an alternative method of earthing in which the neutral of the incoming supply also forms the earth return path. In other words, instead of the neutral and earth of the incoming supply being separate, they are combined. The name protective multiple earthing is usually abbreviated to P.M.E.

The installation within the building is carried out in exactly the same way as for any other system, and separate earth continuity conductors are used. The main earth point at the intake is not, however, connected to a separate earth but to the neutral of the incoming service cable. The Electricity Board affording the supply earths the neutral conductor at a number of points on the distribution network and is responsible for seeing that maximum resistance to earth from any part of the neutral conductor does not exceed a prescribed value.

Because with this system the neutral is relied on as the earth, there must be no fuses, cut outs, circuit breakers or switches anywhere in the neutral. In the United Kingdom an Area Electricity Board may not adopt P.M.E. without the permission of the Secretary of State for the Environment, and stringent requirements are made to ensure that the neutral conductor is adequate to carry earth fault currents, that it is truly kept at earth potential and that it is protected against breaks in continuity. The permission of the Postmaster General is also required. This is because the currents into and through the ground at the points of multiple earthing could cause interference to adjacent telephone and telegraph cables in the ground.

After early hesitations, P.M.E. is becoming increasingly widespread in the United Kingdom. Experience has shown that it is in practice as safe as previously used methods, and it has important advantages. In rural areas it makes it unnecessary for the consumer to have his own earth electrode and therefore removes the risks of earth electrodes in the care of unqualified persons. In urban areas it makes the Electricity Board's distribution network cheaper.

Double insulation

Electrical appliances connected to the permanent wiring of a building, whether through plugs into socket outlets or by means of permanent connections, must themselves be protected against faults. There is need for protection against a fault developing on the appliance itself.

When protection of the permanent wiring depends on earthing, the same principle can be used for the appliance. The previous discussion has assumed this, and has proceeded on the basis that the metal casing of an appliance is effectively connected through the earth pin of the plug to the earth connection in the socket. There is, however, an alternative method of achieving safety of appliances which does not depend on earthing the appliances, and this is known as double insulation. For purposes of exposition appliances of the class known as all insulated may be considered as special cases of double insulation.

Double insulation consists of two separate sets of insulation. The first is the *functional* insulation, which is the ordinary insulation of the conductors needed to confine the current to the conductors and to prevent electrical contact between the conductors and parts not forming part of the circuit. The *supplementary* insulation is additional to and independent of the functional insulation. It is an entirely separate insulation which provides protection against shock in the event of the functional insulation's breaking down.

Another term which we have to explain is *reinforced insulation*. This is an improved functional insulation with such mechanical and electrical properties that it gives the same degree of protection against shock as does double insulation.

An all insulated appliance is one which has the entire enclosure made of substantial and durable insulating material. In effect, it is a double insulated appliance in which the supplementary insulation forms the enclosure.

Both earthing and double insulation provide protection against the breakdown of the primary functional insulation. Earthing depends, in the ways already described, on ensuring that if the functional insulation fails, exposed metalwork will be prevented from rising significantly above earth potential. If the case of the appliance can be made of insulating material which is robust enough to withstand all conditions in which it is to be used, then there is no danger of shock even if the functional insulation fails. Such an appliance is an all insulated one, and this form of construction gives adequate safety, but some appliances must have exposed metal; for example, hedge clippers and portable drills. Other appliances are so large that it is impracticable to make an insulating case strong enough to withstand ordinary usage without making the whole appliance too heavy and cumbersome; for example, a vacuum cleaner. In these cases, double insulation can be used. It provides a second barrier of insulating material between conductors and exposed metal parts. The presence of this additional barrier is a protection against the failure of the functional insulation and makes it unnecessary to earth the exposed metal. Double insulated equipment is designed so that in general two independent sections of insulation must both fail before any exposed metal can become live. The functional and protective insulation must be so arranged that a failure of either is unlikely to spread to the other. They ought, therefore, to be mechanically distinct, so that there is a surface of discontinuity between them.

The principle of double insulation is illustrated in Fig. 104, in which the live and neutral conductors each consist of a cable with ordinary functional insulation. The casing is itself of insulating material and forms the supplementary insulation. If a fault develops on the functional insulation, a short circuit may develop between the live and neutral conductors but the supplementary insulation will prevent the metal handle on the outside from becoming live.

An appliance which is double insulated by the use of either supplementary or reinforced insulation should not be earthed and is not provided with an earth terminal. When it is connected to a standard three pin plug the earth pin of the plug is left unconnected. Double insulation gives the same degree of protection against shocks as earthing, and makes that protection independent of the earth loop impedance. It also gives protection against high impedance faults to earth on the appliance itself, and thus guards against fires caused by local overheating at the appliance.

Double insulation is a means of making an appliance safe. It cannot give protection against faults on the permanent wiring in the fabric of a building. Thus, if there are any metal parts used in the wiring installation, they must still

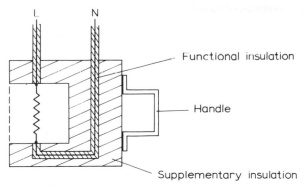

Fig. 104   Double insulation

be earthed. Such parts would be conduit, switches, distribution boards, control panels and so on. It is hardly practicable to carry out an installation of any size without the use of some metal components so that the principle of double insulation cannot be applied to the permanent wiring. Moreover, the designer and erector of the services in a building cannot control what appliances may be connected to the service during the life of the building. Even if all the appliances in a building when it is first put into use are of the double insulated pattern, so long as appliances depending on earth protection (and all electric kettles do) remain in existence, one of them could at some time be connected to the system. Therefore, the system must have an effective earth for the appliances to be linked to.

One of the strongest arguments for the use of double insulated and all insulated appliances is that it makes the safety of the portable appliance independent of the installation it is connected to. The manufacturer knows that the user is protected by the design of the appliance and he has given that safety without having to rely on the earthing system of the building in which the appliance is to be used, over which the appliance manufacturer has, of course, no control. He has thus gone a long way towards protecting the user against the latter's own ignorance of the safe way of connecting an appliance to a defective system, which could happen either through ignorance or through inadvertence.

At the same time, if all appliances were known to be double insulated, the designer of the electrical services could concentrate on protective devices for the system he is designing without having to consider what protection to leave for appliances which the occupants will bring along later. Unfortunately, this is not the case, and the designer must consider the interaction of other people's appliances and his system. In particular, he should perhaps reflect on what kind of flexible cords and extension flexes are likely to bridge the gap between his system and a well made and safe appliance.

The I.E.E. Regulations do not allow the designer to assume that equipment connected to socket outlets will be double insulated and they must therefore be provided with earth connections. Circuits serving double insulated fixed appliances may have the facility for earthing omitted only if the designer is certain that the appliance will never be changed for one of a different type. It is difficult to imagine a situation where a designer can be sufficiently certain of what will happen to his installation in the future to rely on this.

Portable tools

We shall conclude this chapter by considering the special problems of portable tools. These are a very important class of appliances, especially in some factories which use them in large numbers, and they are subject to certain difficulties of their own. When a tool stalls, it is likely to blow a fuse, and after this has happened a few times the operator or maintenance engineer decides that to avoid replacing the fuse every time the tool is momentarily overloaded, he will put in a larger fuse, capable of carrying the overload. Unfortunately, it is also capable of carrying a substantial earth fault current, and the operator can be electrocuted before the oversize fuse clears the fault.

The flexible cables of portable tools come in for exceptionally rough usage and are therefore particularly liable to develop faults. At the same time a man using a portable tool in a factory is likely to be either standing on, or else touching or close to, substantial metal parts, so that he has a low impedance to earth. Thus an earth fault is more probable on a portable tool and its effects are more likely to be immediate and serious than on almost any other kind of appliance. It ought, therefore, to receive extra care in protection but in fact may receive less than average attention.

To overcome these difficulties, factories which use portable tools in large numbers often install a special low voltage supply to serve them. Typical voltages used are 110 V and 50 V. Of course, the tools have to be wound for this supply and it is a disadvantage that the motors are bulkier and heavier. It is perhaps debatable whether it is an advantage or disadvantage that the tools cannot be used outside the factory on a normal supply.

Another solution is to provide protection against sustained overload by means of circuit breakers with time delay characteristics such that they will not operate on temporary overloads, and separate protection against earth faults by earth leakage circuit breakers. A fuse is then needed only for back up protection, if at all, and can be large enough not to blow when the tool temporarily stalls. It is also advisable to use earth monitoring on the earth conductors to portable tools. This makes it necessary for the flexible cables to have an extra conductor and for the tools themselves to have an extra connection.

None of these precautions is very suitable for the home handyman, who may not have enough understanding of electrical theory to appreciate the need for them. It is probably better for portable tools to be protected by double insulation rather than by reliance on effective earthing. A similar case can be argued for making domestic appliances double insulated. Many modern appliances, of which we may quote vacumm cleaners and hair dryers as examples, are nowadays made with double insulation.

Standards relevant to this chapter are:

BS 88   Cartridge fuses for voltage up to and including 1000 V a.c. and 1500 V d.c.
BS 646   Cartridge fuse links for a.c. and d.c.
BS 1361   Cartridge fuses for a.c. circuits in domestic and similar premises
BS 1362   General purpose fuse links for domestic and similar purposes (primarily for use in plugs)
BS 3036   Semi-enclosed electric fuses (ratings up to 100 A and 240 V)
BS 3535   Safety isolating transformers
BS 3871   Miniature and moulded case circuit breakers
BS 4293   Current operated earth leakage circuit breakers
BS 5486   Factory built assemblies of switchgear and control gear

10 Fire Alarms

Introduction

A fire alarm circuit, as its name implies, sounds an alarm in the event of a fire. There can be one or several alarms throughout a building, and there can be several alarm points which activate the warning. The alarm points can be operated manually or automatically; in the latter case they may be sensitive to heat, smoke or ionization. There are clearly many combinations possible, and we shall try in this chapter to give some systematic account of the way they are built up.

Circuits

The simplest scheme is shown in Fig. 105. Several alarm points are connected in parallel, and whenever one of them is actuated the circuit is completed and the alarm sounds. This is described as an open circuit, and it will be seen that it is not fail safe, because if there is a failure of supply, the fire alarm cannot work. Another characteristic of this circuit is that every alarm point must be capable of carrying the full current taken by all the bells or hooters working together.

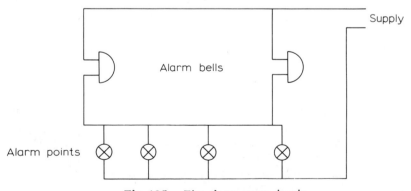

Fig. 105   Fire alarm open circuit

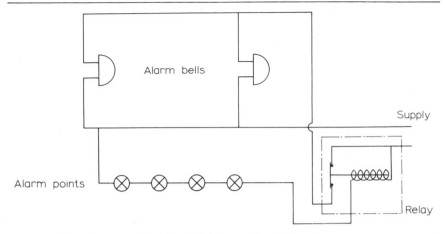

Fig. 106   Fire alarm closed circuit

A slightly more elaborate scheme is shown in Fig. 106. The alarm points are connected in series with each other and with a relay coil. The relay is normally closed when de-energized, and opens when the coil is energized. Thus when an alarm point is activated the relay coil is de-energized, the relay closes and the alarm sounds. This system fails safe to the extent that if the coil circuit fails the main circuit operates the alarm. It is not of course safe against total failure of the supply because in that event there is no supply available to work the bells. The alarm points do not have to carry the operating current of the bells or hooters. This arrangement is called a closed circuit in contrast to the open circuit of Fig. 105. We can notice that in an open circuit the alarm points are wired in parallel and are normally open, whilst in a closed circuit they are wired in series and are normally closed.

When an alarm has been given it is often desirable to silence the audible alarm before the operating point which actuated the alarm is replaced or reset. An alarm stop/reset unit is made commercially which diverts the current from the general alarm to a supervisory buzzer or indicator but restores the current to its normal condition when the alarm initiating point has been reset. An open circuit including this unit is shown in Fig. 107.

In a large building it may be desirable to have an indicator at some central position to show which warning point in the building has caused the alarm to sound. Fig. 108 shows a closed scheme in which each pair of points is connected to a separate signal on an indicator board. The board can have either flags or luminous signs. The circuit can easily be adapted so that each individual point has its own signal or so that a larger number of points is grouped together to one signal. All the points so grouped are wired in series and are connected to their own operating relay in the relay box. The alarm contacts are closed when any one of the relays is energized. The bells can be silenced when required, but

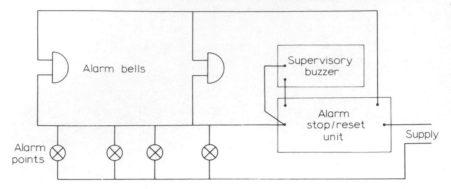

Fig. 107   Fire alarm with relay unit

neither the supervisory buzzer nor the indicator can be reset until the alarm initiating point has been restored to its normal position.

Other refinements can be made for more complicated schemes in large buildings. The exact circuit arrangement must depend largely on the features of the equipment used, and in practice a satisfactory scheme can only be designed round a chosen manufacturer's equipment and with the aid of data from his catalogue.

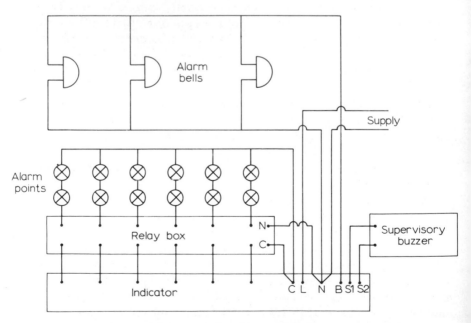

Fig. 108   Fire alarm with indicator

Wiring

The wiring of a fire alarm installation follows exactly the same principles as any other wiring, but greater consideration has to be given to the protection of the wires and to their ability to withstand fire damage.

It is obviously necessary for a fire alarm to go on working for quite some time after a fire has started. The wiring must, therefore, be entirely separate from any other wiring. In conduit or trunking systems, it should be segregated from all other services and run in conduit or trunking of its own. It must be able to withstand high temperature, which in practice means that it is made of either butyl rubber, silicone rubber or MICC cable. The supply to the fire alarm must also be separate from any other supply, and this at the very least means that it must be fed from its own circuit breaker or switch at the main service entry into the building. Some authorities go further and think that the fire alarm system should be at low voltage, and to satisfy this requirement fire alarm equipment is made for 24 V a.c. and 12, 24 or 48 V d.c. operation as well as for mains voltage operation.

A system working on 24 V a.c. has to be fed from a transformer. The primary of the transformer is fed from its own circuit breaker or switch at the main service entry. D.c. systems are fed from batteries of the accumulator type, which are kept charged by a charger unit connected to the mains. As the battery will operate the system for a considerable period before losing all its charge, this method provides a fire alarm which is independent not only of the mains within the building, but also of all electrical services into the building. For this reason, some Factory Inspectors and Fire Prevention Officers insist that a battery system be used. However, when the system voltage is low the currents required to operate the equipment are higher and the voltage drops which can be tolerated are much smaller. We can see this by reflecting that a motor wound for 240 V will work without noticeable diminution of speed or performance if the voltage at its terminals drops by 6 V to 234 V, which is a reduction of 2½%, whereas a 24 V bell may not sound at all if a potential of 18 V is applied to it. In this case a drop of 6 V in the line is a reduction of 25%. At the same time for a given sound output a 24 V bell needs ten times the current that a 240 V bell does.

Thus voltage drop becomes a very serious factor in the design of any low voltage system. The equipment to be used must be carefully checked to see what voltage drop it can accept and the cables sized to keep the drops very low. This usually results in large cables having to be used. There is a great deal to be said for obviating these difficulties as far as possible by not using systems at less than 48 V.

In the author's opinion, there is very little to be said for a 24 V transformer system. It has all the voltage drop problems of a d.c. system without the independence of the incoming service that batteries give. In other words it appears to have the disadvantages of both mains and battery systems without the advantage of either.

The current carrying capacity of each component has to be taken into account in the design and layout of the installation. It may happen, for example, that the total current of all the alarm signals sounding together exceeds the current which can be taken by the contacts of the alarm initiating points. The closed circuit of the type shown in Fig. 106 overcomes this problem because the current to the alarm bells does not go through the initiating points. This is a considerable advantage of the closed system, and can be a deciding factor in choosing it in preference to the open system. Even with closed systems, however, the total current of the alarm signals may exceed the capacity of the contacts in the indicator panel. If this happens, a further relay must be interposed between the indicator and the alarm signals.

Fire alarm points

A typical manually operated fire alarm point is shown in Fig. 109. It is contained in a robust red plastic case with a glass cover. The material is chosen for its fire resisting properties. The case has knock outs for conduit entries at top and bottom but the material can be sufficiently easily cut for the site electrician to make himself an entry in the back if he needs it. Alternative terminals are provided for circuits in which the contacts have to close when the glass is smashed (as in Fig. 105) and for circuits in which the contacts have to open when the glass is smashed (as in Fig. 106). In the former case, there is a test switch which can be reached when the whole front is opened with an Allen Key. In the latter case, the test push is omitted because the circuit is in any case of the fail safe type.

Fig. 109   Fire alarm point (*Courtesy of* Gent & Co. Ltd.)

Fig. 110   Thermal fire alarm (*Courtesy of* Gent & Co. Ltd.)

The alarm point illustrated is suitable for surface mounting. Similar ones are available for flush fixing and in weatherproof versions. The current carrying capacity of the contacts should always be checked with the maker's catalogue.

A thermally operated alarm point is shown in Fig. 110. It consists of a bi-metal strip which deflects when the temperature rises, and thereby tilts a tube half full of mercury. When the tube tilts the mercury flows into the other half of the tube where it completes the circuit between two contacts previously separated by air. Alternatively, the arrangement within the tube can be such that the mercury breaks the circuit when the tube is tilted. The casing of the alarm is of stainless steel. Heat detectors of this type are usually set to operate at 65°C. They are frequently used in boiler houses.

A smoke operated alarm point is shown in Fig. 111. It would be used only in special circumstances which make it necessary to detect smoke rather than heat. This type can cause nuisance operation of the alarm by reacting to small quantities of smoke which have not been caused by a fire; they have for example been known to sound the alarm as a result of cigarette smoke in an office. Modern ones have adjustable sensitivity so that they can be set to avoid nuisance operation.

An ionization detector contains a chamber which houses some low strength radioactive material and a pair of electrodes. The radioactive material makes the air in the chamber conductive so that a small current flows between the electrodes. The size of the current varies with the nature of the gas in the chamber and as soon as any combustion products are added to the air there is a sudden change in the current flowing. The detector also has a second chamber which is permanently sealed so that the current through it never changes. As

Fig. 111   Smoke operated alarm (*Courtesy of* Gent & Co. Ltd.)

long as the currents through the two chambers are equal there is no output; as soon as they become unbalanced there is a net output which is used to operate a transistor switch in the main circuit through the detector.

Bells

Any bell could, of course, be used to sound an alarm. However, the voltage at which it operates must be that of the system and the current consumption must be taken into account in the design of the system. Most manufacturers of fire alarm equipment make bells intended for use with their systems, and it is clearly advantageous to design a system making use of only one maker's equipment. Typical bells have power consumptions of between 1 and 6 W.

Sirens

Electrically operated sirens can be used as an alternative to bells. They are much louder and can be heard at a distance of half to three quarters of a mile. This makes them very suitable for factories in which there may be a lot of background noise. A typical rating is 60 W and we see that a siren takes a much bigger current than a bell. As a consequence, there is a bigger voltage drop in the circuit feeding it. In low voltage circuits, the drop can be sufficiently serious to prevent the sirens from sounding at all, and it becomes especially important to check the layout for voltage drop.

Horns

A horn is an alternative to both bell and siren, and because of its penetrating, raucous note it is particularly suitable where a distinctive sound is needed. Its volume is easily adjustable and its power consumption is intermediate between that of a bell and that of a siren.

The manufacturers of fire alarm equipment also provide standard indicators, relays and reset units for use with the various circuits which we have described in this chapter. Typical indicators are illustrated in Fig. 112. The appropriate

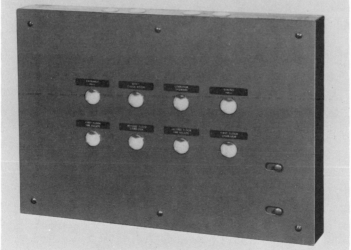

Fig. 112   Fire alarm indicators (*Courtesy of* Gent & Co. Ltd.)

indicator light is illuminated, or a mechanical flag dropped, by a relay which may be either normally energized or normally de-energized. In the first case, the relay holds the signal off and the signal comes on when the activation of the fire alarm point interrupts the circuit. In the second case, the action of the alarm point energizes the relay and brings the signal on. In either case, the indicator must be such that the relay cannot be reset until the alarm point has been returned to its normal position.

Self-contained battery and charger units are also available for low voltage d.c. systems. Alternatively, separate batteries and constant potential chargers can be used. In either case, the designer should check that the ampere-hour capacity of the battery is sufficient for the load it will have to supply. This is of particular importance in those systems which have a current permanently flowing through the alarm initiating points.

Standards relevant to this chapter are:

BS 2740   Simple smoke alarms and alarm metering devices
BS 3116   Automatic fire alarm systems in buildings
BS 5364   Manual call points for electrical fire alarm systems
BS 5446   Components of fire alarm systems for residential premises

I.E.E. Wiring Regulations particularly applicable to this chapter are:

Part   4
Section   531
Section   533
Section   537
Chapter   54
Appendix   3
Appendix   7
Appendix   8

11 Call Systems, Telephone and Public Address Systems

Introduction

In many buildings, it is necessary to have a system of calling staff who are on duty in an office or staff room to go to rooms elsewhere in the building. This happens, for example, in hotels, hospitals and old people's homes. In the case of hospitals, patients wish to call a nurse who is in the ward office to their own beds, and in hotels guests may wish to call staff to their rooms. In old people's homes, a resident may wish to summon either domestic or nursing staff. All such systems can be arranged electrically and form part of the electrical services in a building.

Hotel call systems

Fig. 113 illustrates a scheme used in hotels. There is a push in each room connected to its own relay in a combined relay/indicator unit. One such unit is mounted in a convenient position in the corridor of each floor, and all the room pushes on that floor are wired to it. Each floor unit is in turn connected to an indicator and bell in the manager's office. When the button in a room is pushed, the corresponding relay pulls down the indicator flag for the room in the indicator unit and this in turn pulls down the indicator flag for that floor in the manager's office and at the same time sounds a bell in the office. The circuit has to be reset at the floor indicator unit. The scheme can be varied to have a bell on each floor and an indicator without bell in the main office. Other variations can also be devised to suit individual users' preferences.

Hospital call systems

In hospitals, it is desirable for each patient to be able to call a nurse to his bed. Fig. 114 is a wiring diagram of a simple circuit for achieving this. There is a call unit at each bed which contains a push button, a relay and an illuminated reset lamp push. When the patient pushes the button, the relay is energized and holds

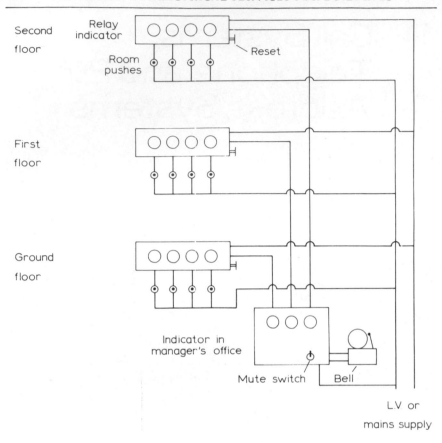

Fig. 113   Hotel call system

itself in until it is released by the reset lamp push. While the relay is energized, a lamp is illuminated; it can be either next to the patient's bed or over the door of the room At the same time, a buzzer and light are operated in the ward office or wherever else the nurse is to be called from. The buzzer can be silenced by a muting switch, but the light can only be cancelled by the resetting push on the patient's unit. Secondary or pilot lamps and buzzers can be placed elsewhere so that a nurse's attention can be attracted in more than one place or so that a Sister can supervise the activity of her staff.

When the call is made, the duty nurse has no indication from where it is coming and must, therefore, walk round all the places on the system until she sees the individual lamp which has been illuminated. The system is therefore limited to small areas. An extended version which does not have such a limitation is shown in Fig. 115. A call unit is provided at each bed and also in each toilet; the toilet unit differs from the bed unit by being operated by

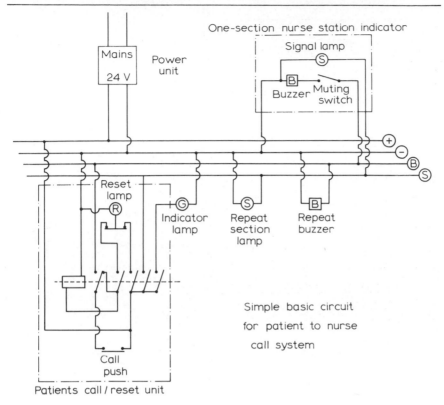

One-section nurse station indicator

Signal lamp

Mains 24 V

Power unit

Buzzer Muting switch

Reset lamp

Indicator lamp

Repeat section lamp

Repeat buzzer

Call push

Patients call / reset unit

Simple basic circuit
for patient to nurse
call system

Fig. 114   Hospital call system

external push buttons instead of by buttons within it. There is a button in each W.C. compartment and in each bath compartment, all of them being wired in parallel with each other to the operating contacts of the single toilet unit. Each call unit illuminates a lamp over the door of the room in which that unit is. Thus all the units in a ward illuminate the same lamp over the ward door. Several rooms are grouped in a section and when any unit in one section is energized a signal lamp for that section is illuminated in the nurse's office. There can also be a parallel signal lamp at a suitable position in the corridor.

When the duty nurse receives a call, the indicator in her office shows which section the call is coming from. When she gets to that section of corridor, she will see which door has a light over it, and once inside the room she can see which call unit has the reset lamp alight. Further central stations can be added in parallel with the main one in case staff have to be called from more than one area or in case supervision of staff activities from another office is necessary.

It will have been noticed that the reset button on the call unit is in the form of a lamp which stays alight until it has been pushed to reset the unit. One

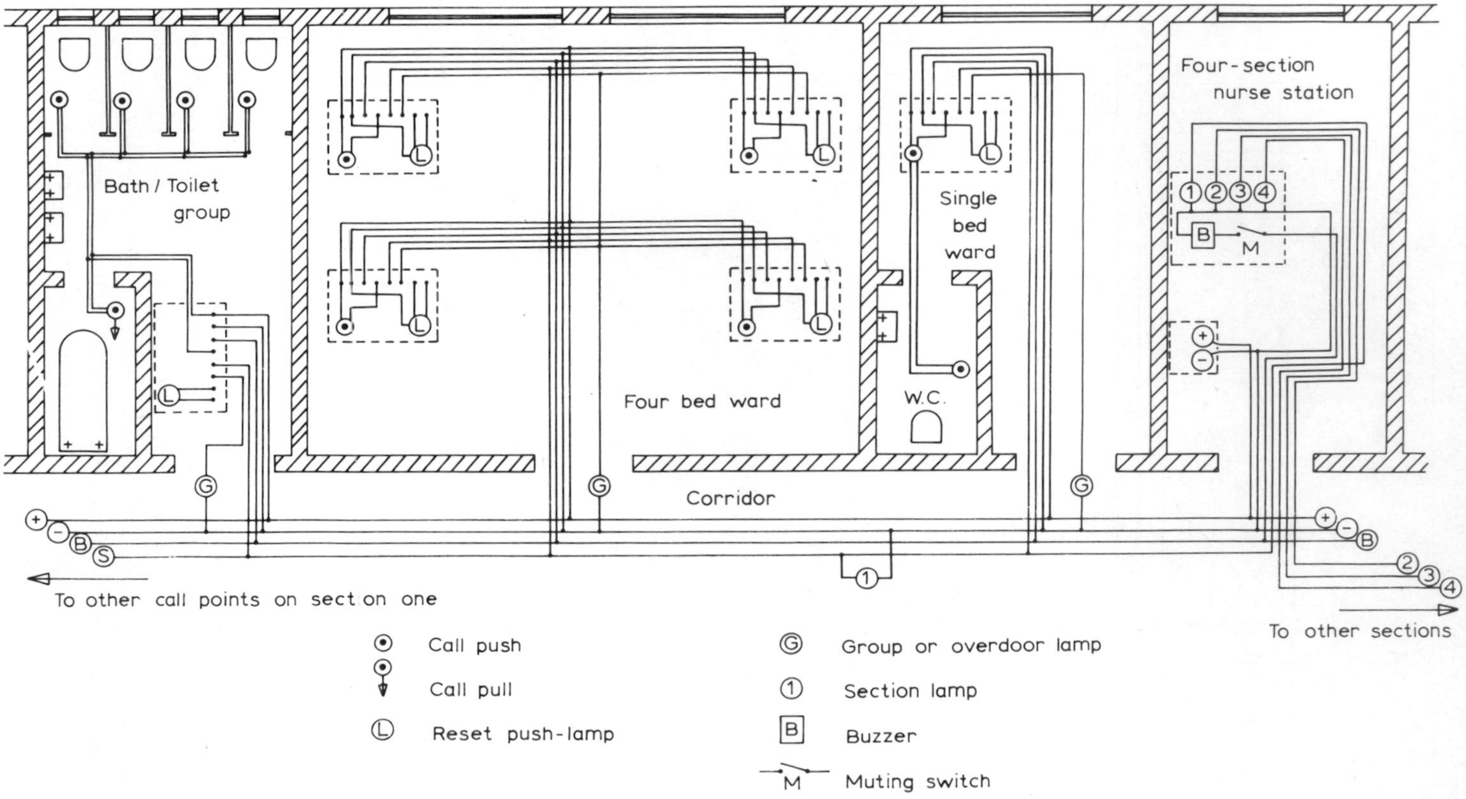

Fig. 115 Hospital call system

Four-section nurse station

Single bed ward

W.C.

Four bed ward

Corridor

Bath / Toilet group

To other call points on sect on one

To other sections

G   Group or overdoor lamp

①   Section lamp

B   Buzzer

M   Muting switch

⊙   Call push

⊙→ Call pull

Ⓛ   Reset push-lamp

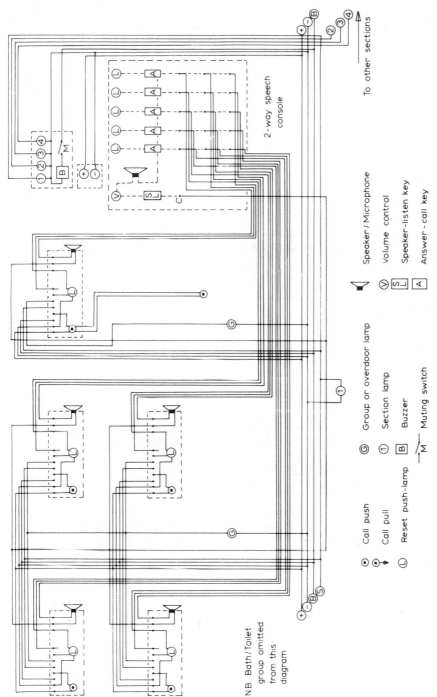

Fig. 116 Call system with speech facility

reason for this is to show which unit in a ward of several beds has been operated, but the presence of the light also helps to reassure a nervous or frightened patient that assistance is on the way. It is simpler and cheaper to make the reset push contain the light than to install a separate button and lamp.

The call units and central station can also incoporate a combined microphone and loudspeaker so that the nurse can speak to the patient from her office. The diagram of Fig. 115 is repeated with this facility added in Fig. 116.

The indicator units and signal lamps are all made for low voltage so that there is no possibility of electrical shock to the users if any faults occur. Another reason for using low voltage is that it would be unnecessarily expensive to have mains voltage signal lamps. It is, therefore, usual to make the whole system a low voltage one and to size the cables accordingly. The design of the wiring thus becomes similar to that for fire alarms discussed in Chapter 10, but it is not so essential to protect the wiring from fire damage. Ordinary PVC cables may be used and there is not the same objection to the use of a transformer from the main building supply. In fact, call system equipment is normally made for a 24 V d.c. supply, and the manufacturers also make a power unit consisting of a transformer and rectifier. 250 V a.c. is supplied to the primary of the power unit and 24 V d.c. is taken from the secondary or output side. The cables must obviously be sized to keep the voltage drop to an absolute minimum, and in a large building it may be necessary to have several independent systems in order that the cables are sufficiently short.

Fig. 117   Alarm indicator (*Courtesy of* Wandsworth Electrical Mfg. Co. Ltd.)

Fig. 118   Indicator lamp (*Courtesy of* Walsall Conduits Ltd.)

A typical indicator to receive the calls in the central office is shown in Fig. 117. It will be seen that this particular unit has three section lights, a buzzer and a buzzer-muting switch. An indicating lamp suitable for mounting over a room door is shown in Fig. 118. It has a cast iron box with a porcelain lampholder to take a 15 W pigmy sign type lamp and an overlapping brass cover with a ruby glass dome. The box is built into the wall so that the cover is flush with the face of the wall.

A single call button is shown in Fig. 119. This is contained in a box and is

Fig. 119   Call button (*Courtesy of* Walsall Conduits Ltd.)

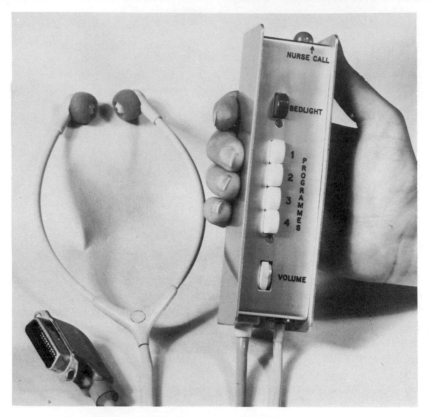

Fig. 120   Call unit (*Courtesy of* Walsall Conduits Ltd.)

suitable for fixing in a wall within reach of a bath or toilet. Examples of complete call units are shown in Fig. 120 and 121. They contain the push button, reset light and also switches for the patient's bedside radio outlet. They are on the end of a wander lead and are designed so that they can be held comfortably in the patient's hand. The wander lead ends in a multiway connector plug which fits into a mating socket in a wall near the bed. The permanent wiring of the installation ends in this socket. The call units can also contain a switch for a bedside lamp and the microphone for a speech system.

Another version of a call unit, which is suitable for fixing to a bedside locker, is shown in Fig. 122.

Door indicators

A call system which is being increasingly installed is the door indicator. This is an illuminated sign, with or without a buzzer, fixed outside an office door and operated by a push at the desk in the office. The sign can be 'Enter', 'Engaged',

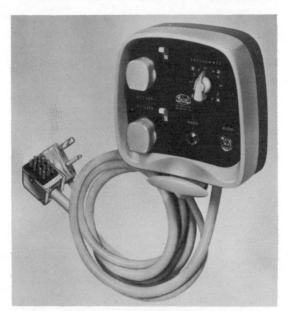

Fig. 121 Call unit (*Courtesy of* Walsall Conduits Ltd.)

Fig. 122 Call unit

'Next Patient' or any other message required. An enter sign is shown in Fig. 123, together with its wiring diagram. A transformer is contained in the same box as the sign and the system operates at low voltage. The table push energizes a relay, which is also contained in the sign box, and this connects the lamp behind the sign to the low voltage supply. The equipment can be arranged so that the light stays on until reset by the table push. Standard equipment for these applications is made by a number of manufacturers. The particular one shown is suitable for surface mounting, but they are also made for flush fixing, and clearly for a new building this would be preferable.

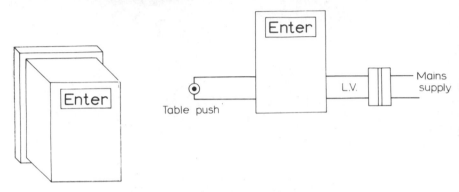

Fig. 123   Door sign

Telephone systems

The design of telephone systems is beyond the scope of this book, but we must consider the provision that has to be made for them within a building. In many cases all that is needed is a route by which the public telephone service, which in the United Kingdom is British Telecom, can bring a telephone cable to an instrument. British Telecom telephones are operated by batteries at the Telephone Exchanges and need no source of power within the buildings they serve. Telephone cable is quite small and if the position of the outlet for the telephone receiver is known it is sufficient to install a 20 mm conduit from outside the building to the outlet, with the same number and spacing of draw in points as are used for any other conduit system It is usual for the electrical installer to fix the conduit and leave draw wire in it, which the telephone engineers subsequently use for pulling their cable in after the building is finished and occupied.

The most common procedure used in the United Kingdom is for British Telecom to supply plastic ducts which the builder puts in the ground from the telephone main in the road to a point just within the building, and for the electrical contractor to supply and install metal conduit from the end of the plastic duct to the final telephone outlet position. A conduit box is provided where the plastic duct meets the metal conduit, and the electrical contractor puts draw wire into both the duct and the conduit.

If the telephone cable is to come in overhead, as is likely to happen in rural areas, then British Telecom will do all the outside work, including the fixing of a terminal on the wall of the building. The electrical installer then has only to provide conduit with draw wire from the entry point, which in this case will be at high level, to the final telephone outlet position.

Some buildings have an internal telephone system which may consist of extensions to the public telephones or may be an entirely separate installation. Here again the essential matter for the electrical services designer is to agree the outlet positions with his customer and to arrange for them to be linked to each other by conduit or trunking. Trunking can be a useful alternative to conduit when the system is a complex one needing many cables with a large number of junctions. Telephone cables do not have a protective sheathing and therefore need the mechanical protection of conduit or trunking. They can however be run exposed on surfaces and this is often done, but in a new building it makes it rather apparent that the designers forgot about the telephones until it was too late.

An internal telephone installation which is independent of the public telephones must receive power from somewhere. All telephones work on low voltage and this is provided either by a battery or by an electronic power pack. A battery needs to be kept charged by a battery charger which in turn has to be supplied by mains power. A power pack usually contains its own transformer but this must then be fed from the mains. In whichever way the telephone works, mains power has to be provided somewhere, usually at the central exchange of the system. The power required is very small and can be supplied from a socket outlet or fused spur unit on the nearest convenient general purpose power circuit.

Most British Telecom telephones take their power from batteries at the Telephone Exchange, and do not need power from the consumer's supply. Some of the British Telecom private branch exchange systems do however include a power pack which requires a supply from the subscriber's premises. The electrical designer should therefore discuss the system to be used with British Telecom and make sure that any necessary power outlets are provided. They can again be ordinary socket outlets or fused spur units taken from a convenient power circuit at a point adjacent to the telephone equipment.

Public address systems

Public address and loudspeaker systems are somewhat similar to telephones. The details of the equipment to be used can be settled only with manufacturers' catalogues and by discussion with the manufacturers. Once the equipment and its location have been selected provision must be made for running cables from the announcing station to the loudspeakers. Because these cables are likely to be put in after all other building work is finished a conduit system is the almost inevitable choice. Loudspeaker cables are like telephone cables in that they are small and do not have an outer sheath; this also makes it difficult to find any alternative to putting them inside conduit.

Standards relevant to this chapter are:

BS 5613   Alarm systems for the elderly and others living at risk

12 Low Voltage Systems

The use of low voltage wiring for fire alarms and call systems has been discussed in the previous two chapters. Other applications occur in laboratories, where permanently installed low voltage outlets are required for various experiments. Permanent outlets are easier for the staff than the use of accumulator batteries which have to be carried from preparation room stores and set up on the laboratory benches for each experiment. Low voltage supplies are also needed for microscopes which have a built-in light for illuminating the slide.

The requirements for any particular laboratory must, of course, be agreed by the electrical designer with the staff who will be using the laboratory. In some cases, as for example illuminating microscope slides, the supply is needed at a constant voltage. Such a supply can be provided by a transformer. Usually, this is comparatively small and can be mounted on a bracket fixed to the wall either of the laboratory or of an adjacent store. For a given power, the current on a low voltage service is higher than on a mains voltage circuit and, therefore, the cable sizes soon become substantial. To prevent the use of excessively large cables, it is convenient to keep down the number of outlets on one circuit and to use a separate transformer for each secondary circuit.

It would be possible to take the secondary of the transformer to a low voltage distribution board and split there to several low voltage circuits. The cable from the transformer to the distribution board would, however, be very large to take the necessary current, and it is better to use a separate transformer for each secondary circuit. The kVA rating required is calculated from the secondary voltage and total output power needed. As usual in this kind of design, it is advisable to allow ample spare capacity so that the transformer rating should be somewhat above the calculated requirement.

The voltage on the primary of the transformer is known, being the ordinary mains supply voltage in the building, and this determines the transformer ratio. The ratio determines the primary current and thus provides all the information necessary to design the mains circuit feeding the transformer. A fuse can, if desired, be provided on the secondary of the transformer, but an overload on the secondary would draw an overload on the primary so that the fuse in the supply to the transformer will also protect the secondary. This will not, however, be the case if the primary has been oversized while the secondary has not. The exact carrying capacities of the primary and secondary sides should be carefully compared before the fusing arrangements are finally decided on.

In other cases, a laboratory needs a variable low voltage output. This makes it possible to set up experiments with a choice of voltage. Low voltage laboratory units are made which contain a transformer, a rectifier and variable tappings on the output switches. The one illustrated in Fig. 124 gives a choice of outputs ranging from 6 V to 24 V a.c. and from 6 V to 24 V d.c. It can, therefore, be set to give the output required for whatever experiment is in hand.

The wiring arrangements are exactly the same as for a fixed output transformer. The cables must, of course, be sized for the largest current which

Fig. 124   Low voltage unit (*Courtesy of* Chloride Industrial Batteries Ltd.)

will be taken, which will correspond to the lowest voltage. It is also at the lowest voltage that voltage drop along the cable is most serious.

A number of accessories is available for terminating the low voltage wiring at the benches, and some of these have been described in Chapter 1. For laboratory work, the type shown in Fig. 20 (page 22) is the most suitable because the wires used for setting up the bench experiment can be readily connected to these terminals. Microscopes or other permanent equipment requiring a low voltage supply will usually have a trailing flexible lead with a plug at the end of it. For these, a socket outlet is clearly more convenient than a laboratory type of terminal, but it should not be possible to plug a low voltage piece of equipment into a mains voltage socket. Low voltage wiring for such applications should, therefore, terminate in 5 A two-pin socket outlets which should be clearly labelled with the voltage. Two-pin plugs can then be supplied and fixed to the equipment, and it will not be possible for any one to push these into ordinary socket outlets.

Intrinsically safe circuits

The use of flameproof equipment is discussed in Chapter 1. An alternative approach in areas where there is a risk of fire is the use of intrinsically safe circuits. The principle of these is that the energy of any spark which occurs shall be limited so that it is not sufficient to ignite the vapour. This is achieved by using low voltages and equipment which does not take high currents at these voltages. The power in the circuit is thus low and there is not enough energy available to initiate combustion.

Not all equipment can be designed on this basis, but it is often possible to have intrinsically safe circuits within a hazardous area operating relays which control normal equipment outside the danger zone. This may be cheaper than installing flameproof equipment within the zone.

Standards relevant to this chapter are:

BS 1259   Intrinsically safe electrical apparatus and circuits
BS 1538   Intrinsically safe transformers for bell signalling circuits

I.E.E. Wiring Regulations particularly applicable to this chapter are:

Section   411
Regulation   553—3

13 Communal TV Systems

Introduction

If every tenant in a block of flats had his own TV aerial on the roof of the building, the result would be very unsightly. In the case of a 24 storey tower block having four flats on each floor, the roof would probably not be large enough to accommodate 96 separate aerials. Even a low rise development looks ugly if every house has its own aerial; sometimes a four storey development consists of two flats and a maisonette above each other so that each 'house' would need three aerials. In fringe reception areas, one needs large aerials mounted very high up. This makes it difficult to equip each dwelling with its own aerial. In built up areas, large buildings shield smaller ones, so that if an estate consists of a number of small blocks and one or two towers, occupants of the small blocks left to provide their own aerials might find it desirable to put them on masts rising as high as the top of the tower block.

For these reasons, it is an advantage to receive television and radio signals at one suitably sited aerial array and relay them to individual dwellings by cables or transmission lines. Very large relay systems exist, serving whole towns, sometimes from a mast receiver several miles away. The building services engineer is more likely to be concerned with community systems serving a single block of flats or one small estate of houses and maisonettes. In this book, we confine our attention to such community systems.

Radio signals are electro-magnetic waves in space. They cover a range of frequencies which are classified as shown in Table 2. Since the product of frequency and wavelength is always equal to the velocity of light, which is constant, the frequency is inversely proportional to the wavelength. Radio broadcasts have in the past usually been identified by the wavelength, but at the frequencies used for television the wavelength becomes so small that it is more convenient to use the frequency. The properties of aerials and transmission lines depend very much on frequency and different types have to be used for

Table 2   Frequency bands

Designation	Abbreviation	Frequency range
Low frequency	LF	30 kHz — 300 kHz
Medium frequency	MF	300 kHz — 3 MHz
High frequency	HF	3 MHz— 30 MHz
Very high frequency	VHF	30 MHz— 300 MHz
Ultra high frequency	UHF	300 MHz— 3 000 MHz
Super high frequency	SHF	3 000 MHz—30 000 MHz

different frequencies. It is convenient to subdivide the VHF and UHF frequencies into five bands, which makes it possible for commercial equipment to be manufactured for one or two selected bands only. This subdivision is shown in Table 3.

If a signal consists of one frequency only, the only way it can convey information is by varying in amplitude. As soon as several adjacent frequencies are present, they can combine to form complicated waveshapes and the total number of distinguishable patterns increases rapidly. This is a simplified explanation of why the band of frequencies required for a transmission increases as the amount of information to be conveyed increases. Television provides much more information than sound broadcasting, and therefore each service requires a large bandwidth. For sound broadcasting, each station needs a bandwidth of only 10 kHz. A station broadcasting on 1500 m (which corresponds to 200 kHz) actually uses all wavelengths between 1457 m and 1543 m. Provided the next station has a nominal wavelength of 1587 m or more there will be no interference between them; it is obvious that there is no difficulty about keeping stations separate from each other under these conditions.

A 625 line TV picture, on the other hand, requires a bandwidth of 5.5 MHz. It is immediately obvious from Table 2 that this cannot be transmitted at less

Table 3   Broadcasting services

Range	Band	Channel numbers	Frequency	Service
LF	—	—	150— 285 kHz	AM sound long wave
MF	—	—	535—1605 kHz	AM sound medium wave
HF	—	—	2.3— 26.1 MHz	AM sound short wave
VHF	I	1— 5	41— 68 MHz	TV band I
	II	—	87.5— 100 MHz	FM sound (VHF)
	III	6—13	175— 215 MHz	TV band III
UHF	IV	21—34	470— 582 MHz	TV band IV
	V	39—68	614— 854 MHz	TV band V

than HF, and that the HF range would only accommodate five different stations. This is the reason that TV is transmitted in the VHF and UHF ranges. Even within these, care has to be taken about separation of stations. The five bands of frequency are, therefore, further divided into a number of channels each of which covers a bandwidth of 6 MHz. These channels are also indicated in Table 3. Each station is allocated one channel, and neighbouring stations are thus prevented from interfering with each other.

Since the distance over which VHF and UHF waves can be propagated is quite limited two stations more than a certain minimum geographical distance apart can safely use the same channel.

Aerials

If an e.m f. is placed in the centre of a short wire (Fig. 125), the two halves of the wire act as capacitor plates, one becoming positively charged and the other negatively. Each charge produces an electric field. Suppose now that the e.m.f. is alternating; there is then an alternating charging current in the wire. When the current is a maximum the positive and negative charges occupy the same place and produce equal and opposite fields. When the current is zero the positive and negative charges are at opposite ends of the wire and produce a resultant electric field. Thus there is an electric field which alternates with the charging current in the wire.

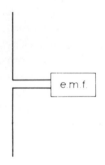

Fig. 125   Principle of dipole

The current also produces a magnetic field which spreads out from the wire with the velocity of light. The motion of this magnetic field induces a further electric field.

Now the oscillating charges in the wire have not only a velocity, but also an acceleration. This acceleration is propagated outwards in the electric field at a finite velocity and, therefore, the field further out is moving with a lower velocity than that closer in. Since the charges oscillate the acceleration is alternately forward and backward, and the result is that the complete field radiated forms closed loops which travel out from the wire and expand.

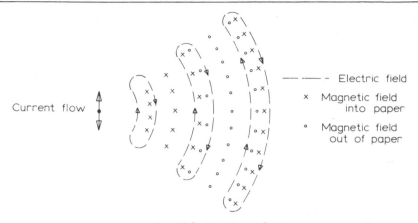

Fig. 126   Radiated field

(Fig. 126). It can be shown that this radiated field is appreciable only if the length of the wire is of the same order as the wavelength.

The total electric field thus contains three terms. The first two are the induction fields and the third is the radiation field. Near the aerial the induction fields predominate, but they become negligible at a distance greater than about five wavelengths. At larger distances, the radiation field is the only important one.

By a converse mechanism, a wire placed in an alternating electric field and suitably orientated to that field will have an e.m.f. induced in it. In fact, it turns out that receiving aerials are identical to transmitting aerials. In practice, receiving aerials can be made simpler than transmitting aerials because high efficiency is not so important when receiving as when transmitting.

The simple aerial of the type shown in Fig. 125 is known as a dipole and its total length is half the wavelength radiated or received. The description given above is an attempt to explain in simple physical terms how a dipole radiates and receives. Any electric current is associated with a magnetic and an electric field, the relationship always satisfying Maxwell's equations. The field radiated by an aerial is thus the solution of Maxwell's equations for the boundary conditions given by the current distribution in the aerial and the geometry of the aerial. The mathematical difficulties of calculating such solutions are so great that theoretical solutions are not given in even the most advanced textbooks on radio propagation, and aerials are designed on the result of experimental investigations. The general principle remains that the dimensions of the aerial are approximately equal to the wavelength. If follows from this that different aerials are required for different frequency bands.

Aerials are used for services ranging from telegraphy through navigation, broadcasting and telephony to radar and space communications. There is a very large number of types of aerial in use to cover the wide range of applications

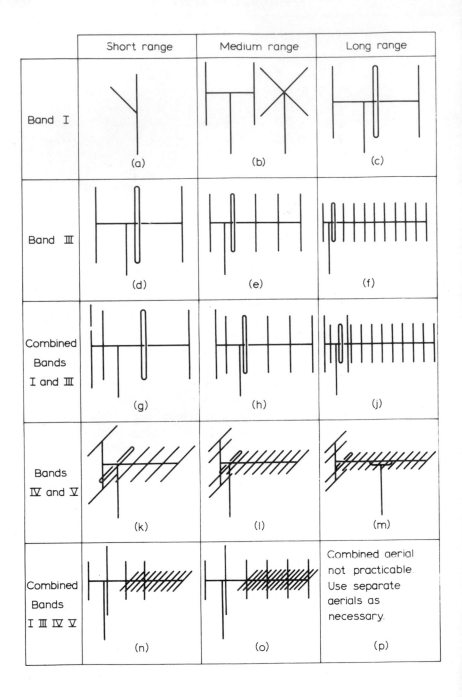

Fig. 127 TV aerials

and frequencies employed, but the following are the types most likely to be encountered by the building services engineer. They are all simple variations of the basic dipole described above.

For band I, a single dipole as shown in Fig. 127a can be used for reception at a short range from the transmitter. If the receiver is some distance from the transmitter a modification is introduced to increase the strength of the dipole. It has been found that the addition of elements not connected to the receiver cable can strengthen the signal at the dipole. An element slightly longer than the dipole and spaced about a quarter of a wavelength behind it has the effect of reflecting waves onto the dipole and thus strengthening the signal to it. This is illustrated in Fig. 128 and results in the well known H aerial shown in Fig. 127b. A modification of this is the X aerial, also shown in Fig. 127b. The principle is illustrated in Fig. 129 from which it is seen that the dipole and reflector are partially folded on themselves. It has been found that they then operate like a

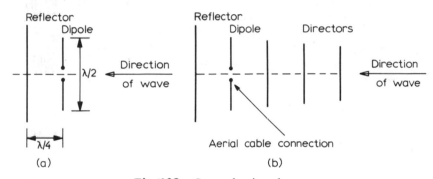

Fig. 128   Strengthening elements

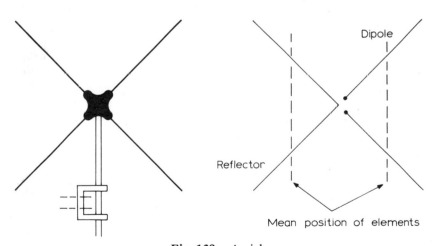

Fig. 129   Aerial

straight pair of elements occupying the mean positions shown dotted in Fig. 129.

Further improvement in reception can be obtained by adding directors in front of the dipole at distances of about a quarter of a wavelength, and progressively shorter than the dipole (Fig. 128). It is found that the effect of these is to increase the sensitivity of the dipole to radiation from the direction in which the array is pointing.

An arrangement such as that of Fig. 128 is known as a Yagi array.

For long range reception on band I, an aerial consisting of dipole, reflector and director (Fig. 127c) is used.

As explained later in this chapter, maximum power is transferred from a source to a load when their impedances are equal. The half wave dipole has an impedance of about 75 ohms, which matches the characteristic impedance of the co-axial aerial cable used. When other elements, that is to say reflectors and directors, are added the impedance of the aerial is reduced, and much of the power received is lost at the mismatch between aerial and cable. This can be overcome by folding the dipole, as shown in Fig. 130. It is still half a wavelength long but its impedance is higher, and when it is used in an array its impedance matches that of the cable. Yagi arrays, therefore, generally use a folded dipole.

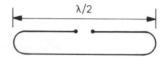

Fig. 130   Folded dipole

Similar aerials are used for the reception of FM radio on band II, but they are put horizontally instead of vertically. This is simply because the radiation broadcast for this service is polarized in the horizontal plane whereas that for television services is polarized in the vertical plane. The same types of aerials are used for band III, but because of the shorter wavelengths the strength of the signal decreases more rapidly and, therefore, a larger number of elements is used in the array. Typical arrays are shown in Fig. 127d, e and f, and combined band I and band II aerials are shown in Fig. 127g, h and j.

At UHF the range of transmitters becomes much less, although there is less interference between neighbouring stations. Therefore, the receiving aerials have a larger number of elements. These are readily accommodated because as a result of the smaller wavelength they are shorter and more closely spaced. In fact a six-element UHF array can present a neater and more compact appearance than a single dipole for band I. Also as a consequence of the shorter wavelength, the reflector can more effectively take the form of a square mesh. Typical arrays are shown in Fig. 127k, l and m

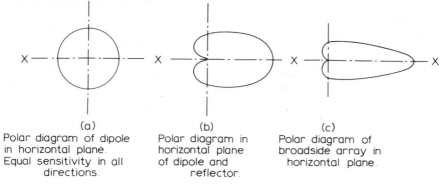

(a)
Polar diagram of dipole
in horizontal plane.
Equal sensitivity in all
directions.

(b)
Polar diagram in
horizontal plane
of dipole and
reflector.

(c)
Polar diagram of
broadside array in
horizontal plane.

In all cases the complete polar diagram
is the solid of revolution about X·—·X

Fig. 131 Polar diagrams

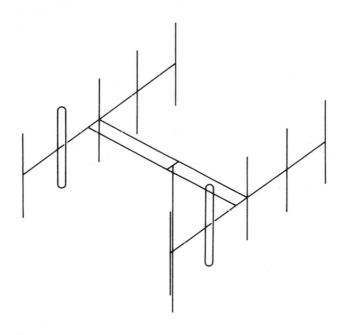

Fig. 132 Twin arrays

All the aerials we have described are directional. The voltage induced in them depends on the angle between the axis of the array and the plane of the wave of radiation. It is a maximum when the axis of the array is perpendicular to the wavefront and is zero when the axis of the array is parallel to the wavefront. The sensitivity can be represented by the length of a line drawn in the direction of the advancing wavefront. The locus of the ends of all such lines forms the polar diagram. This is illustrated in Fig. 131.

Greater sensitivity in the axial direction can be obtained by altering the shape of the polar diagram as shown in Fig. 131c. One way of achieving this is to have a pair of Yagi arrays mounted side by side in a broadside arrangement, as shown in Fig. 132. It will be appreciated that in all cases the axis of the aerial array must point as closely as possible towards the transmitter.

Transmission lines

The signal received by the aerial is sent from the aerial to the television outlets along an aerial cable or television transmission line. The design of the line is an important part of the design of a communal TV system, and we must learn something about transmission engineering to understand it. Transmission engineers are very much concerned with loss of power, which is usually measured in decibels. A decibel is one-tenth of a bel, and a bel is the logarithm to base 10 of the ratio of two powers. If the power at the sending end is P_s and that at the receiving end is P_r the loss of the line is

$$\log_{10}\left(\frac{P_s}{P_r}\right) = 10 \log_{10}\left(\frac{P_s}{P_r}\right) \text{ decibels.}$$

The decibel is convenient because of the very large losses and amplifications encountered in communications engineering; for example, if an amplifier has an output 10 000 times the input it is more convenient to say that it has a gain of 40dB. Since the gain, or loss, in decibels is a ratio, the input level should also be stated.

Power ratios are proportional to the square of the voltage or current ratios. Therefore:

$$\log_{10}\left(\frac{P_s}{P_r}\right) = 2 \log_{10}\left(\frac{V_s}{V_r}\right) = 2 \log_{10}\left(\frac{I_s}{I_r}\right)$$

Thus when measurements are made in volts or amperes the loss is

$$20 \log_{10}\left(\frac{V_s}{V_r}\right) \text{dB} \quad \text{or} \quad 20 \log_{10}\left(\frac{I_s}{I_r}\right) \text{dB.}$$

If the logarithms to base 'e' are used instead of to base 10, the unit is the neper instead of the bel. This is not used so often, but it is sometimes convenient because the attenuation of a transmission line per unit length is always a power of 'e'.

There is a loss of power in all transmission lines. At very low frequencies this is largely due to the resistance of the line, although inductance and capacitance are important for exact calculation of long power lines. At high frequencies inductance and capacitance become much more important, and it can readily be shown that the losses increase rapidly with frequency. At frequencies above 3000 MHz the losses in cables are so high that transmission by cables is no longer possible and the only way of conveying energy at these frequencies is by waveguides.

The power engineer distributing power at 50 Hz is interested in supplying the power taken by the load with the minimum loss in the line. The communications engineer sending signals to a receiver has a rather different outlook. He is handling much smaller quantities of energy, and he can install amplifiers along the line which feed in energy from an independent source without distorting the wave shape of the signal. His main concern is to provide a strong enough signal to the receiver at the end of the line. It can be shown that maximum power is taken by a load when that power equals the power lost in the generator. Although at this condition maximum power is taken by the load, the efficiency of the transmission is only 50%. This is clearly uneconomic for power transmission but is practicable in telecommunications where the magnitude of the signal is much more important than transmission efficiency. The difference between the operating points of power and communications systems is shown in Fig. 133.

When an alternating voltage is applied to the sending end of an infinite line, a finite current flows because of the capacitance and leakage inductance between the two wires forming the line. The ratio of voltage applied to current flowing is the input impedance. The input impedance for an infinite length of line is known as the *characteristic impedance* and is denoted by Z_0. It should be noted that the characteristic impedance varies with frequency.

Since the line is infinite, no current waves reach the far end. Therefore, there is no reflection and no reflected waves return to the sending end. For the same reason, the current flowing depends only on the characteristic impedance (Z_0) and is not affected by the terminating or load impedance (Z_r) at the far end. Although an infinite line is obviously a purely hypothetical object, in practice the state of affairs we have just described is approximately fulfilled by many long lines. Furthermore, a short line terminating in a load impedance equal to the characteristic impedance of the line (i.e. $Z_r = Z_0$) behaves electrically as if it were an infinite line.

The characteristic impedance is the ratio of voltage to current at any point in an infinite line or in a correctly terminated line. However, the current and voltage are not the same at all points; because of the ordinary impedance of the line, they become progressively less along the line.

Let I_s = current at sending end
        I_1 = current one kilometre down line.

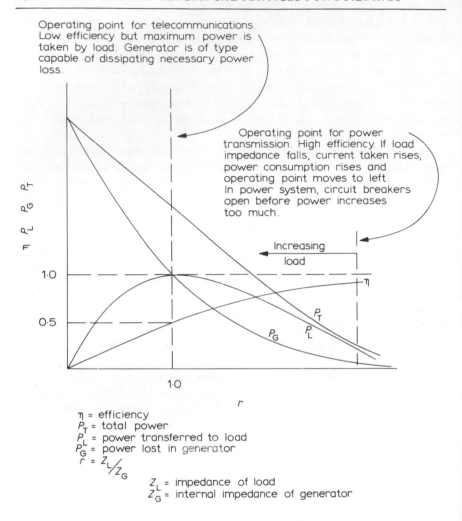

Operating point for telecommunications. Low efficiency but maximum power is taken by load. Generator is of type capable of dissipating necessary power loss.

Operating point for power transmission. High efficiency. If load impedance falls, current taken rises, power consumption rises and operating point moves to left. In power system, circuit breakers open before power increases too much.

Increasing load

η = efficiency
P_T = total power
P_L = power transferred to load
P_G = power lost in generator
$r = Z_L/Z_G$

Z_L = impedance of load
Z_G = internal impedance of generator

Fig. 133   Operating points of transmission lines

Then $\dfrac{I_1}{I_s} = e^{-\gamma}$ where γ = propagation constant per kilometre of line.

γ is a complex quantity so that I is both less than I_s and also different in phase. In general,

$$I_n = I_s e^{-n\gamma}$$ where I_n = current n kilometres along the line.
Similarly, $E_n = E_s e^{-n\gamma}$

γ is a complex quantity which can be written $\gamma = \alpha + j\beta$ where α is the attenuation constant and β is the phase constant.

The four quantities:

Z_0 = characteristic impedance
γ = propagation constant
α = attenuation constant
β = phase constant.

are characteristic of the particular cable being used. They are known as the secondary line constants and can be calculated theoretically from the four primary line constants which are

R = resistance per kilometre (ohms)
G = leakage per kilometre (mhos)
L = inductance per kilometre (henries)
C = capacitance per kilometre (farads)

Whilst the primary constants are independent of frequency the secondary constants in general vary with frequency.

Now a communication signal carries information in its waveshape. It is most important to preserve this shape while the signal is being sent along the line. There are three main causes of distortion:

1. Characteristic impedance varies with frequency. If the line is terminated in an impedance that does not vary with frequency in the same manner as that of the line, distortion will result.

2. Attenuation varies with frequency.

3. The velocity at which the wave shape travels along the line varies with frequency, so that waves of different frequencies arrive at different times.

It can be shown that the condition for minimum attenuation is $LG = CR$. It can also be shown that this condition makes Z_0 independent of frequency. This is called the distortionless condition. On some communication lines the inductance is artificially increased to bring the line nearer to this condition, but this is not found necessary on communal TV systems.

If a line of characteristic impedance Z_0 is joined to an impedance having a value other than Z_0 part of the wave travelling down the line will be reflected back at the point of discontinuity. The reflection is a maximum when the line is either open circuited or short circuited ($Z_r = \infty$ or $Z_r = 0$) and is zero when $Z_r = Z_0$. The current in the line is always the sum of the incident wave and the reflected wave.

This way of looking at matters will seem unfamiliar to power engineers. The rigorous mathematics of transmission lines is the same for power lines as for communication lines, but differences arise from differences in the frequencies at which they are operated. At 50 Hz a line 500 km long is less than one-tenth of a wavelength. Although the stationary distribution of current along it is mathematically equal to the sum of two waves travelling in opposite directions, this fact is of only academic interest to the power engineer. At 3000 MHz, however, the wavelength is only a few metres and the two travelling waves have a physical meaning which is easily visualized.

At the sending end, the current can be represented by a vector of length $I_{s_{max}}$ rotating at an angular frequency ω where $\omega = 2\pi f$ and f is the frequency of the applied voltage. At a point x down the line the current I_x differs from I_s because of the attentuation of the cable and can be represented by a vector of length $I_{x_{max}} = I_{s_{max}} \cdot e^{-\alpha x}$ at an angle of βx to the original vector, but still rotating at an angular frequency of ω. The instantaneous current along the line at successive intervals is then as shown in Fig. 134a. For simplicity in drawing this figure α has been taken as zero, so that the current is the same in magnitude along the line. It will be seen that at all times the envelope of instantaneous current along the line is a sine wave. Moreover, the sine wave moves down the line. The wavelength is that length at which $\beta x = 2\pi$

$$\therefore \quad \beta\lambda = 2\pi$$

$$\therefore \quad \lambda = \frac{2\pi}{\beta}$$

Since   velocity = frequency x wavelength

$$v = f\lambda$$

$$= f \cdot \frac{2\pi}{\beta} = \frac{2\pi f}{\beta}$$

$$= \omega\beta$$

This is the velocity at which the signal travels down the line. It is not the same as the velocity at which energy is transferred.

A wave travelling in the opposite direction will be as shown in Fig. 134b. When the incident and reflected waves are equal in magnitude they combine as shown in Fig. 134c. We can see that there is a wave on the line of frequency f and wavelength $\lambda = \omega/\beta$ but it does not travel along the line. Whereas with the travelling wave the rotating vector has the same magnitude along the line and varies only in phase along the line, in this case the vector changes in magnitude along the line. As a result there are nodes where the current is always zero and anti-nodes where there is a maximum. This type of wave is called a standing wave, and is produced by the combination of equal incident and reflected waves. If only part of the incident wave is reflected then there is a standing wave with a travelling wave superimposed on it. Power is transferred only by the forward travelling wave, so that standing waves represent a loss. Reflected waves also originate at junctions where one cable branches into two or three.

Complete reflection occurs when a line is either open circuited or short circuited. Under these conditions, there is no forward travelling wave, and no

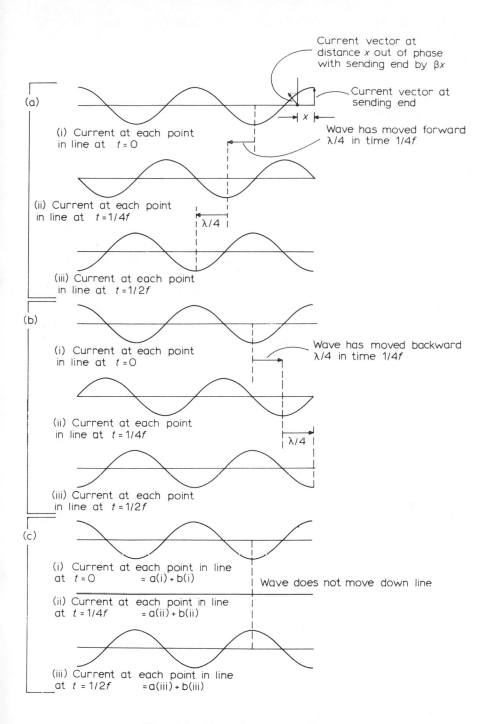

(a)

(i) Current at each point in line at $t = 0$

Current vector at distance x out of phase with sending end by βx

Current vector at sending end

Wave has moved forward $\lambda/4$ in time $1/4f$

(ii) Current at each point in line at $t = 1/4f$

$\lambda/4$

(iii) Current at each point in line at $t = 1/2f$

(b)

(i) Current at each point in line at $t = 0$

Wave has moved backward $\lambda/4$ in time $1/4f$

(ii) Current at each point in line at $t = 1/4f$

$\lambda/4$

(iii) Current at each point in line at $t = 1/2f$

(c)

(i) Current at each point in line at $t = 0$ = a(i) + b(i)

Wave does not move down line

(ii) Current at each point in line at $t = 1/4f$ = a(ii) + b(ii)

(iii) Current at each point in line at $t = 1/2f$ = a(iii) + b(iii)

Fig. 134   Travelling and standing waves

power is tranferred. This is readily understandable, because neither an open circuited nor a short circuited line feeds a load.

For maximum transfer of power the internal impedance of the generator must be matched to the characteristic impedance of the line and this must in turn be matched to the input impedance of the load. In the case of the communal TV system, the generator is the aerial. If the matching is not correct, standing waves are formed in the line, and power is dissipated in the line. Some of this dissipation takes the form of radiation which causes interference to neighbouring aerials.

In general, the impedance of the generator is not the same as that of the line, and that of the line is not the same as that of the receiver. Therefore, at each of these points of discontinuity some form of impedance transformer is required. Such a transformer can be made from a four pole network which has different input impedances between the two pairs of terminals. In Fig. 135, the line with characteristic impedance Z_1 sees the impedance Z_1 between terminals a and b and is therefore, correctly terminated. The load with impedance Z_r sees itself supplied from an impedance Z_r between terminals c and d and is, therefore, also correctly matched. There is a small loss of power in the four pole network, but this is preferable to the large losses that would occur in the line if the mismatching were permitted to remain.

A similar problem has to be solved where there is a branch. In Fig. 136a, the

Fig. 135   Impedance transformer

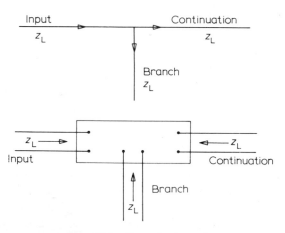

Fig. 136   Branch network

input cable sees the continuation and branch cables in parallel and is in effect terminated by an impedance of $Z_1/2$. A network has to be provided as in Fig. 136b, so that each of the three cables sees an impedance of Z_1. There are a number of devices which have the necessary characteristics, but many of them are sensitive to frequency and operate correctly only over a limited band of frequencies. In communal TV practice, a simple arrangement of resistances is used, as shown in Fig. 137. These junction boxes also perform another function which we shall come to later in this chapter.

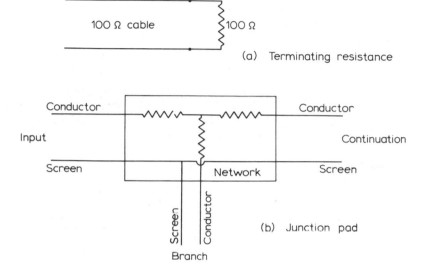

Fig. 137   Junction box

Cables

Having discussed the theory of a transmission line, we are now in a position to say something about the cable which actually forms the line. We start by noting that any wire carrying current tends to act as a radiating aerial. At low frequencies the power radiated from an ordinary wire is so small that it can hardly be detected, but at high frequencies it can become significant. Not only is there a loss of power from the line itself, but the radiation will cause interference in neighbouring receivers. Similarly, any wire acts as a receiving aerial, and a line feeding a television or radio set can pick up unwanted high frequency radiations. Both these effects can be suppressed by efficient screening, and radio and TV services therefore always use screened cable.

A line may be a single conductor using the earth as a return. This has capacitance to the earth and is termed an unbalanced line. If a conductor is provided for the return, it can be arranged so that either the two conductors have different capacitances to earth or that they have the same capacitance to

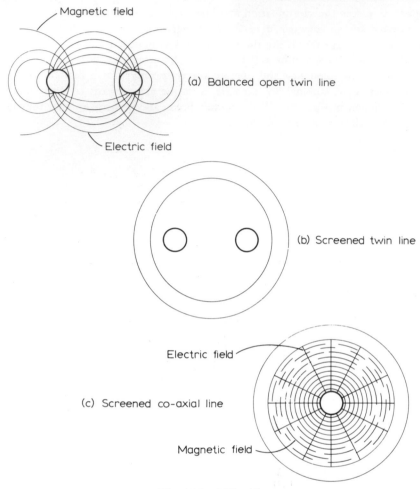

(a) Balanced open twin line

(b) Screened twin line

(c) Screened co-axial line

Fig. 138   TV cables

earth. The former arrangement is termed unbalanced and the latter balanced. Typical arrangements are shown in Fig. 138.

The energy conveyed by a transmission line is in fact held in the electric and magnetic fields associated with the current and voltage. In the case of open lines, these fields extend infinitely into space. At high frequencies the energy is rapidly dissipated into space, and the losses from the transmission system become unbearably high. If a screen is placed round the conductors, the fields are confined within the screen and the losses are reduced. Typical forms of radio frequency cable are shown in Fig. 139. The cable of Fig. 139a has an inner conductor of copper wire and an outer conductor of seamless lead tube. It is suitable for high frequency transmission and high power aerial feeders. Fig. 139b

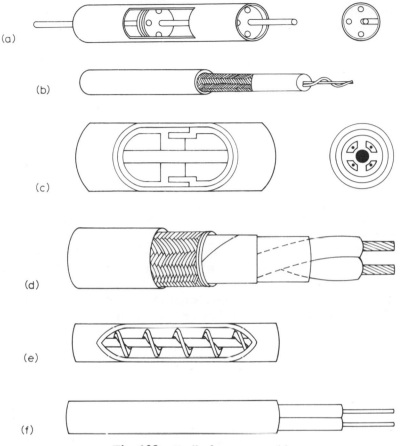

Fig. 139   Radio frequency cables

shows a cable with the inner conductors supported in the centre of a tube of polyethylene and an outer conductor of wire braid with a PVC or lead alloy sheath. Fig. 139d is a screened and balanced twin feeder. Fig. 139e shows a co-axial cable relying mainly on air as the insulation between the conductors with an insulating helical thread supporting the inner conductor. Fig. 139f is similar to Fig. 139d, but does not have the two conductors wound over each other; it is often used in radio relay systems.

As the frequency increases the losses in the dielectric become more important than the resistance loss, and for high frequency cables the properties of the dielectric must be carefully considered. The best performance can be obtained by air spaced cables, but cables with good solid dielectrics are used where economics outweigh purely technical considerations. Communal TV systems normally employ screened co-axial cable with polyethylene dielectric.

Frequency translation

In communications, information is contained in a complete waveform covering a band of frequencies. This band of frequencies is evenly spaced about the carrier frequency, but the wave can be transferred to any other band of frequencies of equal width, without any loss of information. It can be transmitted in this form and later transferred back to its original frequency band, or the information can be read out in the new band.

The losses in a line increase with frequency and in the early days of communal TV systems it was impracticable to send signals along cables at UHF frequencies because the losses were too high. UHF broadcasts were therefore translated near the aerial masthead to suitable channels in the VHF range and transmitted along the communal distribution system in this form. They were not translated back at the receiving points and the receiving sets had to be adjusted accordingly. The tuner in a TV set contains a number of pre-wired tuning or filter circuits, each set to a particular channel. The station selector switch connects the appropriate filter, leaving the others out of circuit. All that was necessary therefore was for a service technician to take out one pre-wired circuit and replace it with another pre-wired to the new channel. This was a very simple operation.

The frequency changing equipment accepted the broadcast signals and translated them to the required channels by reference to high stability local oscillators.

Modern amplifiers have made it possible to send UHF signals along aerial distribution cables so that frequency translation is no longer necessary, but it may still be encountered on some existing systems.

Mixers and splitters

Because the signal received by the aerial is attenuated as it travels along the cable, it must be amplified. There are difficulties in designing amplifiers which work equally well over a large range of frequencies and, therefore, two or more amplifiers are used, each operating on a particular band of frequencies. The output impedance of each amplifier must be matched to the characteristic impedance of the cable. Also the output of one amplifier must not feed back into another amplifier to distort the output of that one. It is, therefore, necessary to insert a mixer unit between the amplifiers and the line. The mixer unit has to accept two or more different frequencies and combine them, but at the same time isolate their sources one from another. It achieves this by suitable filtering networks of inductances and capacitances.

The layout of a scheme sometimes makes it necessary to take two cables away from one amplifier. The output impedance of one amplifier must then be matched to the characteristic impedances of two cables working in parallel. This is done by a splitter unit which divides the output from an amplifier and distributes it between two or more lines. The splitter is a network of resistances, inductances and capacitances chosen according to the conditions under which the division is to be made.

Power loss and amplification

As we have already said, because of losses in the transmission system, the signal received at the aerial has to be amplified either at the aerial or along the line or both. Now as the gain of an amplifier is increased, the noise it introduces also increases, and this sets a limit to the gain which can be used. In practice, amplifiers with a gain of 30 to 60 dB are used. If a 30 dB amplifier is used, then the distribution system can be allowed to attentuate the signal by 30 dB before a repeater amplifier has to be installed. Similarly a 60 dB amplifier permits losses of 60 dB to be incurred before a repeater is necessary.

Attenuation occurs at a uniform rate along the length of the cable, but at each branch there is a sharp loss in the junction unit. Consequently, the graph of signal strength against cable run appears as in Fig. 140. It will be seen that the signal level at each branch decreases as one goes along the cable. A TV set must receive a signal not less than about 1 mV but will distort the picture if the signal is more than about 6 dB higher than this minimum. The signal level at a junction must be high enough to accommodate the losses in the length of line continuing

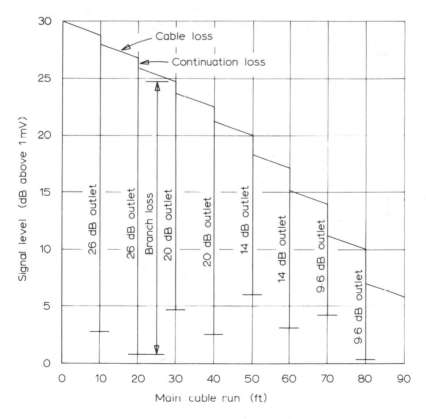

Fig. 140   Attenuation graph

from the junction to the next amplifier. The attenuation in the shortest branch from the junction must be large enough to bring the signal strength down from that at the junction to less than the maximum acceptable to the receiving set at the end of the short branch. The branch cable is quite short and in any case its length cannot be adjusted to yield the required attenuation. It is, therefore, necessary to build in some extra attenuation, and this is done in the junction unit itself. The junction unit attenuates the signal to the branch outlet terminals by a given amount whilst keeping the attenuation to the line continuation terminals as low as possible. This is the second function of the junction unit which we referred to above and it is achieved by a suitable network of resistors.

It will be seen from Fig. 140, that the attenuation required to produce a given output signal level is different at each junction. It would be most inconvenient to make a special unit for every junction, but fortunately this is not necessary. A good TV set has a certain tolerance in the input voltage it can accept, so that a standard attenuator can be used for several successive junctions giving a small range of outputs within the limits acceptable to the receivers. Table 4 shows a standard range of ratios which have in practice been found adequate in a large number of cases. The resulting signals available at the outlets in a typical case are also shown in Fig. 140.

Table 4   Standard junction attenuators

Reduction of outlet signal relative to input		Reduction of continuation signal relative to input	
Ratio	dB	dB	Ratio
100 to 1	40	0.1	1.01 to 1
50 to 1	34	0.2	1.02 to 1
20 to 1	26	0.46	1.05 to 1
10 to 1	20	0.9	1.11 to 1
5 to 1	14	1.9	1.25 to 1
3 to 1	9.6	3.5	1.50 to 1

Typical systems

It is generally found that up to about 50 dwellings can be served from one repeater amplifier. Two typical schemes are shown in Figs 141 and 142.

Fig. 141 indicates a housing development consisting of two blocks of dwellings. Each block has 17 single storey flats on the ground floor (intended for old people) and three layers of maisonettes above them. Each of these layers consists of three floor levels; the entrance to all maisonettes is on the middle layer, and alternate maisonettes have the bedrooms below and above the entrance and living rooms. Access corridors thus occur only on the ground floor and floors 3, 6 and 9, the other floors containing rooms reached by internal stairs within the maisonettes. All services follow the same distribution pattern,

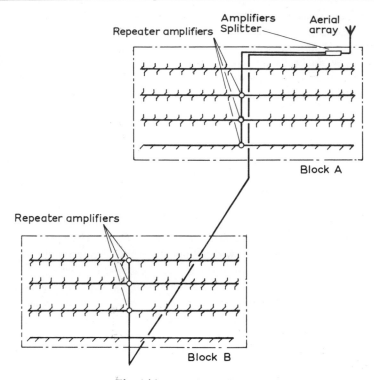

Fig. 141   Typical scheme

that is to say, they run in the ceiling of the access corridors and rise and drop into alternate maisonettes.

There is an aerial array which can receive three existing television channels and which also has provision for the reception of future services on three other channels. The array is mounted on the tank room on the roof of one of the blocks. The receiving equipment is fixed inside the tank room and consists of amplifiers and splitters. Two cables are taken from this main station, one to serve each block. They run along a duct in the roof and then drop in a duct alongside the main stairs. One cable drops to the ground and then continues inside a 50 mm plastic conduit under an open space to the other block where it rises in a duct alongside the main stairs. The other cable drops in the same duct of the first block, but has a junction box at level 9. From this two branches run along the ceilings of the access corridors with further junction boxes outside each front door. A third branch from the junction box continues down the duct and feeds repeater amplifiers at levels 6, 3 and 1. From each of these, outgoing cables run along the ceilings of the access corridors feeding junction boxes outside each flat.

The cable entering the second block serves that block in an identical manner except that it works from the bottom up instead of from the top down. The repeaters are therefore at levels 3, 6 and 9, whilst the ground floor is served directly from the main mast head amplifier.

Figure 142 illustrates an estate consisting of one 24 storey tower block and eighteen low blocks. Of the low blocks, numbers 1 to 8 and 17 to 18 are built on top of a podium covering a ground level car park. They are of two storeys and alternate blocks contain maisonettes and a pair of single floor flats. Blocks 9 to 16 start at ground level and are four storeys high. Here alternate blocks contain one maisonette and two flats above each other and two maisonettes above each other. Each floor of the tower block has four flats.

The aerial array is on the tank room of the tower block. The receiving equipment is just inside the tank room and consists of amplifiers and splitters together with a power unit. Eight of the outgoing cables cross the roof and drop inside conduit in the corners of the tower block. In each corner one cable serves the living rooms in the upper half of the block with a junction box at each level from which a short stub cable leads to the aerial outlet. The other cable in each corner drops past these levels without junctions and then serves the living rooms in that corner in the lower half of the block in a similar manner.

Two other cables from the receiving equipment drop in trunking in the central service duct of the tower block and then continue underground in 25 mm polythene conduit. One of them runs along blocks 8 to 2 receiving amplification at points in blocks 8, 9 and 5 and terminating in a final amplifier in block 2. From each of these amplifiers a final outlet cable runs at high level in the car park under each block. Under each living room there is a branch going to the living room above.

From the amplifier at block 5 there is another branch taking the main cable to further amplifiers in blocks 10, 12 and 13. From each of these there is a cable running within a polythene conduit outside the block. There is a junction box in the wall of each bay of these blocks from which an aerial cable runs to each living room outlet. Where there are two or three living rooms above each other two or three branches come off next to each other and run in separate 20 mm conduit to the several outlets. This arrangement makes it unnecessary to enter the lower flat if the cable to one of the upper flats has to be renewed.

The other cable from the tower block goes to block 17 where it feeds an amplifier. The cable branches at this amplifier; one branch goes to amplifiers in blocks 17 and 18 and the other to an amplifier in block 16. Blocks 17 and 18 are fed from their amplifiers in the same way as blocks 1 to 8 and block 16 is fed in the same way as blocks 10 to 13.

On the whole scheme all the junction units are contained in conduit boxes accessible from outside so that any repairs or replacements can be done without technicians having to get into flats. This is an important consideration because it is always difficult to get workmen to a job at a time when all the tenants are there to let them in.

The power to the amplifiers is supplied from the receiving power unit and is

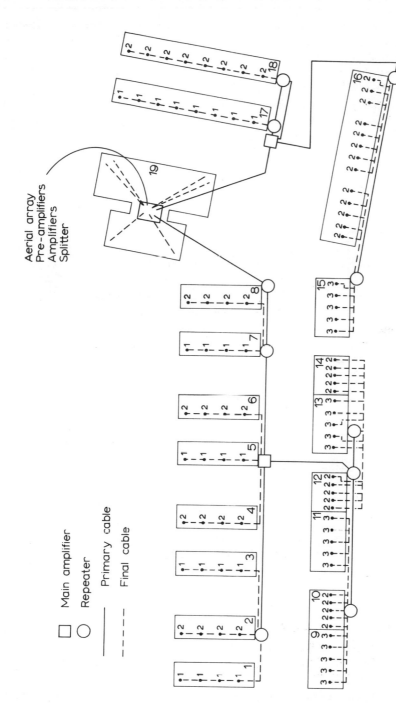

Fig. 142 UHF scheme

fed at mains frequency along the aerial cable itself. This method of line-feeding the amplifiers makes it unnecessary to provide power points at each amplifier position and this results in a significant saving in cost.

A smaller scheme requiring no repeater amplifiers is shown in Fig. 143. This development consisted of a four-storey block A of four maisonettes and four blocks of terraced town houses B, C, D and E. The aerial was mounted on the roof of block A and there was an amplifier with a splitter and a power unit in a cupboard at high level on the common staircase of this block. Four cables ran through 20 mm conduit within this block to serve the four living rooms in it.

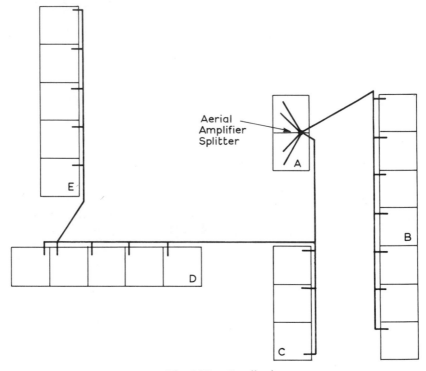

Fig. 143   Small scheme

Two other cables dropped in 20 mm conduit in the block and continued in 20 mm polythene conduit in the ground outside. One ran along blocks C, D and E whilst the other ran along block B. In both cases there was a junction box in the wall of each house from which a short stub aerial cable ran in conduit to the outlet in the living room. The longest cable on this scheme served only 13 dwellings and it was therefore possible to avoid the use of repeater amplifiers altogether. On a small scheme it is better to have a splitter at the masthead with several distributing cables than to run a single cable round the whole site with several repeater amplifiers along it.

Standards relevant to this chapter are:

BS CP 327   Telecommunication facilities in buildings
BS 3041   Radio frequency connectors
BS CP 1020   The reception of sound and television broadcasting

14 Lightning Protection

Lightning strokes can be of two kinds. In the first, a charged cloud induces a charge of opposite sign in nearby tall objects, such as towers, chimneys and trees. The electrostatic stress at the upper ends of these objects is sufficiently great to ionize the air in the immediate neighbourhood, which lowers the resistance of the path between the cloud and the object. Ultimately, the resistance is lowered sufficiently for a disruptive discharge to occur between them. This type of discharge is characterized by the time taken to produce it, and by the fact that it usually strikes against the highest and most pointed object in the area.

The second kind of stroke is a discharge which occurs suddenly when a potential difference between a cloud and the earth is established almost instantly. It is generally induced by a previous stroke of the first kind; thus if a stroke of this kind takes place between clouds 1 and 2 (Fig. 144), cloud 3 may be suddenly left with a greater potential gradient immediately adjacent to it than the air can withstand, and a stroke to earth suddenly occurs. This type of stroke occurs suddenly and is not necessarily directed to tall sharp objects like the first kind of stroke. It may miss tall objects and strike the ground nearby. Fig. 145 shows other ways in which this kind of stroke may be induced. In each case, A is a stroke of the first kind and B is the second type of stroke induced by A. In each case the first stroke from cloud 1 changes the potential gradient at cloud 2 and thus produces the second stroke.

The current in a discharge is uni-directional and consists of impulses with very steep wave fronts. The equivalent frequency of these impulses varies from 10 kHz to 100 kHz. While some lightning discharges consist of a single stroke, others consist of a series of strokes following each other along the same path in rapid succession. The current in a single stroke can vary from a few hundred amperes to a maximum of about 200 000 A, with a statistical average of 20 000 A. It rises to a peak value in a few microseconds. When a discharge consists of several successive strokes, each stroke rises and falls in a time and to

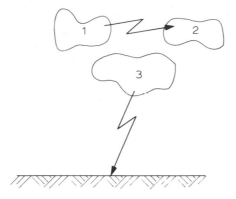

Fig. 144   Induced lightning stroke

an amplitude of this order so that the whole discharge can last a second or even longer.

The effects of a discharge on a structure are electrical, thermal and mechanical.

As the current passes through the structure to earth it produces a voltage drop which momentarily raises the potential of part of the structure to a high value above earth. One function of a lightning conductor is to keep this potential as low as possible by providing a very low resistance path to earth. It is recommended in the British Code of Practice (BS CP 326:1965) that the resistance to earth of the protective system should not exceed 10 ohms. The sharp wave front of the discharge is equivalent to a high frequency current and, therefore, there is also an inductive voltage drop which has to be added vectorially to the resistive drop. Part of the lightning conductor is thus inevitably raised to a high potential. This brings with it a risk of flashover from the

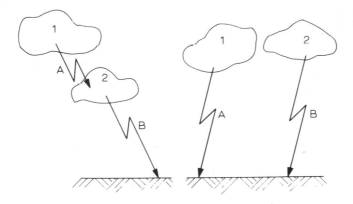

Fig. 145   Induced lightning strokes

conductor to other metal in the structure, such as water and gas pipes and electrical cables. These in turn would then be raised to high potential which could bring danger to occupants of the building, and it is necessary to guard against such flashovers. The discharge of the lightning stroke to earth can also produce a high potential gradient around the earthing electrode, which can be lethal to people and to animals. The resistance to earth of each earthing electrode should be kept as low as is practicable.

The duration of a lightning discharge is so short that its thermal effect can in practice be ignored.

When a large current of high frequency flows through a conductor which is close to another conductor, large mechanical forces are produced. A lightning conductor must, therefore, be very securely fixed.

A lightning conductor works by diverting to itself a stroke which might otherwise strike part of the building being protected. The zone of protection is the space within which a lightning conductor provides protection by attracting the stroke to itself. It has been found that a single vertical conductor attracts to itself strokes of average or above average intensity which in the absence of the conductor would have struck the ground within a circle having its centre at the conductor and a radius equal to twice the height of the conductor. For weaker than average discharges the protected area becomes smaller. For practical design it is therefore assumed that statistically satisfactory protection can be given to a zone consisting of a cone with its apex at the top of the vertical conductor and a base of radius equal to the height of the conductor. This is illustrated in Fig. 146.

A horizontal conductor can be regarded as a series of apexes coalesced into a line, and the zone of protection thus becomes a tent-like space (Fig. 147). When there are several parallel horizontal conductors the area between them has been found by experience to be better protected than one would expect from the

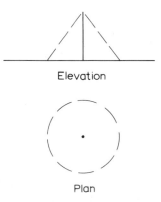

Elevation

Plan

Fig. 146   Protected zone

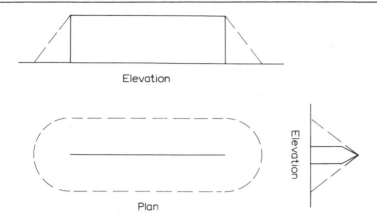

Fig. 147 Protected zone — horizontal conductor

above considerations only. On the basis of experience the recommended design criterion is that no part of the roof should be more than 9 m from the nearest horizontal conductor except that an additional 0.3 m may be added for each 0.3 m by which the part to be protected is below the nearest conductor.

Whether or not a building needs protection against lightning is a matter of judgement. It obviously depends on the risk of a lightning stroke and also on the consequence of a stroke. Thus a higher risk of a strike can probably be accepted for an isolated small bungalow than for, say, a children's hospital. Whilst no exact rules can be laid down which would eliminate the designer's judgement entirely some steps can be taken to objectify the assessment of risk and of the magnitude of the consequences. BS CP 326:1965 gives seven tables (reproduced in Table 5) of index figures for use of structure, type of construction, consequential effects, degree of isolation, type of country, height of structure and lightning prevalence. In each table an index number between 1 and 10 is allocated to each of a number of categories. The index figures are added together to give a total risk index. The higher the risk index the greater the need for protection, and the code suggests that protection should be provided if the risk index is more than about 40. A map showing the number of thunderstorm days in Great Britain, which is needed for use with the last table, is reproduced in Fig. 148. These extracts from BS CP 326 are reproduced here by permission of the British Standards Institution (2 Park Street, London W.1) and the reader who wishes to study the topic further should obtain a copy of the Code of Practice from the Institution.

A complete lightning protective system consists of an air termination network, a down conductor and an earth termination. The air termination network is that part which is intended to intercept lightning discharges. It consists of vertical and horizontal conductors arranged to protect the required

Table 5   Need for lightning protection

Index A:	*Use of structure*	*Index*
	Houses and similar buildings	2
	Houses and similar buildings with outside aerial	4
	Factories, workshops, laboratories	6
	Offices, hotels, blocks of flats	7
	Places of assembly, churches, halls, theatres, museums, department-stores, post offices, stations, airports, stadiums	8
	Schools, hospitals, children's and other homes	10
Index B:	*Type of construction*	
	Steel framed encased with non-metal roof	1
	Reinforced concrete with non-metal roof	2
	Brick, plain concrete, or masonry with non-metal roof	4
	Steel framed encased or reinforced concrete with metal roof	5
	Timber framed or clad with roof other than metal or thatch	7
	Brick, plain concrete masonry, timber framed, with metal roof	8
	Any building with a thatched roof	10
Index C:	*Contents or effects*	
	Contents or type of building	
	Ordinary domestic or factory building, factories and workshops not containing valuable materials	2
	Industrial and agricultural buildings with specially susceptible contents	5
	Power stations, gas works, telephone exchanges, radio stations	6
	Industrial key plants, ancient monuments, historic buildings, museums, art galleries	8
	Schools, hospitals, children's and other homes, places of assembly	10
Index D:	*Degree of isolation*	
	Structure in a large area of structures or trees of same height or greater height, e.g. town or forest	2
	Structure in area with few other structures or trees of similar height	5
	Structure completely isolated or twice the height of surrounding structures or trees	10
Index E:	*Type of country*	
	Flat country at any level	2
	Hill country	6
	Mountain country between 300 m and 1000 m	8
	Mountain country above 1000 m	10

Index F:	*Height of structure*			
	Up to 9 m	2	24—30 m	11
	9—15 m	4	30—38 m	16
	15—18 m	5	38—46 m	22
	18—24 m	8	46—53 m	30
	Higher structures require protection in all cases			
Index G:	*Lightning prevalence*			
	No. of thunderstorm days per year			
	Up to 3	2	13—15	14
	4— 6	5	16—18	17
	7— 9	8	19—21	20
	10—12	11	over 21	21

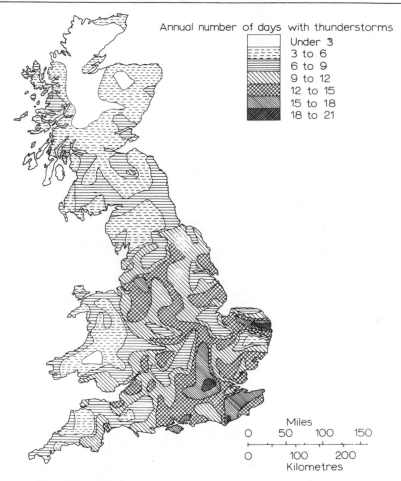

Annual number of days with thunderstorms
Under 3
3 to 6
6 to 9
9 to 12
12 to 15
15 to 18
18 to 21

Miles
0   50   100   150

0   100   200
Kilometres

Fig. 148   Map showing annual number of thunderstorm days

area in accordance with the empirical rules which we have given above. Typical arrangements are shown in Fig. 149.

The earth termination is that part which discharges the current into the general mass of the earth. In other words, it is one or more earth electrodes. These have already been discussed in Chapter 9; earth electrodes for lightning protection are no different from earth electrodes for short circuit protection systems. The total resistance of an electrode for a lightning protection system should not exceed 10 ohms, but it is better if it is even less. The electrodes should be the rod or strip type, and should be either beneath or as near as possible to, the building being protected. Plate electrodes are expensive and come into their own only when large current carrying capacity is important. Because of the short duration of a lightning stroke, this is not a consideration for

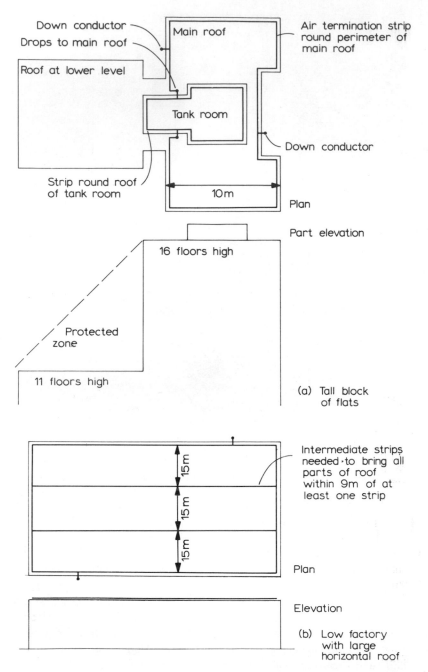

Down conductor

Drops to main roof

Roof at lower level

Main roof

Air termination strip round perimeter of main roof

Tank room

Down conductor

Strip round roof of tank room

10 m

Plan

Part elevation

16 floors high

Protected zone

11 floors high

(a) Tall block of flats

15 m

15 m

15 m

Intermediate strips needed to bring all parts of roof within 9m of at least one strip

Plan

Elevation

(b) Low factory with large horizontal roof

Fig. 149

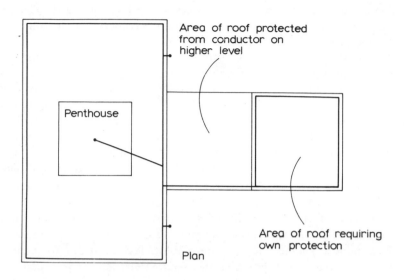

Area of roof protected from conductor on higher level

Penthouse

Plan

Area of roof requiring own protection

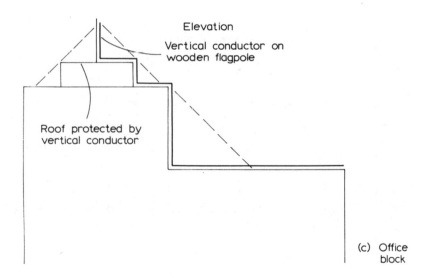

Elevation

Vertical conductor on wooden flagpole

Roof protected by vertical conductor

(c) Office block

Typical lightning conductors

lightning electrodes. The practice sometimes adopted of putting the electrode some distance away from the building is both unnecessary and uneconomical, and may increase the danger of voltage gradients in the ground.

The down conductor is the conductor which runs from the air termination to the earth termination. A building with a base area not more than 100 m^2 needs only one down conductor. For a larger building, there should be one down conductor plus a further one for every 300 m^2 in excess of the first 100 m^2. Alternatively for a larger building one can have one down conductor for every 30 m of perimeter. The number chosen can be the smaller of the numbers given by these alternative methods of calculation. For buildings higher than 30 m a single down conductor is in general enough. A tall non-conducting chimney should, however, have two down conductors equally spaced, with a metal conductor round the top of the chimney joining the two down conductors. The down conductors should preferably be distributed round the outside walls of the building. If this is for any reason not practicable a down conductor can be contained inside a non-metallic and non-combustible duct. It can, for example, run inside a service duct provided the service duct does not contain any non-metal-sheathed cables. Sharp bends, as for example at the edge of a roof, do not matter, but re-entrant loops can be dangerous. A re-entrant loop produces a high inductive voltage drop which can cause the lightning discharge to jump across the loop. The discharge can, for example, go through the masonry of a parapet rather than round it. On the basis of experience it can be said that this danger may arise when the perimeter of the loop is more than eight times the length of the open side. This is illustrated in Fig. 150. If a parapet is very narrow the problem can be solved by taking the conductor through a hole in the parapet as shown in Fig. 151.

Sometimes a building is cantilevered out at a level above the ground. If the

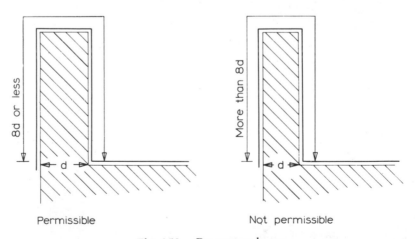

Permissible   Not permissible

Fig. 150   Re-entrant loops

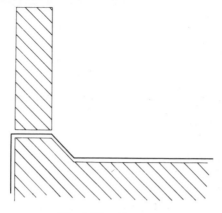

Fig. 151   Parapet

down conductor followed the contour of the building, there would be a real risk of flashover under the overhang, which could be lethal to anyone standing there. In such a building the down conductor must be taken straight down inside the ducts within the building. This problem and its solution are illustrated in Fig. 152.

The material used for lightning conductors is normally aluminium or copper. The criterion for design is to keep the resistance from air termination to earth to a minimum. Since the bulk of the resistance is likely to occur at the earth electrode the resistance, and therefore the size, of the down conductor would not appear to be critical. Recommended dimensions are given in Table 6. Larger conductors should be used if the system is unlikely to get regular inspection and maintenance.

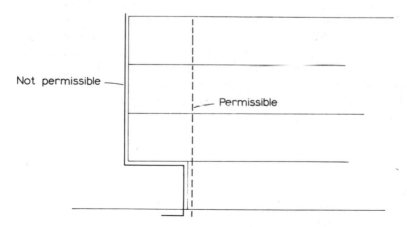

Fig. 152   Cantilevered building

Table 6   Lightning conductors

Component	Minimum dimensions
Air terminations	*mm*
Aluminium and copper strip	20 x 3
Aluminium, aluminium alloy, copper and phosphor bronze rods	10 diam.
Stranded aluminium conductors	19/2.50
Stranded copper conductors	19/1.80
Down conductors	
Aluminium and copper strip	20 x 3
Aluminium, aluminium alloy and copper rods	10 diam.
Earth terminations	
Hard drawn copper rods for driving into soft ground	12 diam.
Hard drawn or annealed copper rods for indirect driving or laying in ground	10 diam.
Phosphor bronze for hard ground	12 diam.
Copper clad steel for hard ground	10 diam.

External metal on a building should be bonded to the lightning conductor with bonds at least as large as the conductor.

When a lightning conductor carries a stroke to earth, it is temporarily raised to a considerable potential above earth. There is, therefore, a risk that the discharge will flashover to nearby metal and cause damage to the intervening structure or occupants. This can be prevented either by providing sufficient clearance between conductor and other metal or by bonding them to ensure that there can be no potential difference between them. To calculate the clearance required to withstand the potential difference due to resistance we can assume a current of 150 kA and an electric breakdown strength of 500 kVm^{-1}. This is a figure for air, and the value for solid structural material is higher. To allow for the potential difference due to inductance we can assume a steeply rising current of 4×10^{10} As^{-1}, which is well on the high side of the average likely to occur, and an inductance of the down conductor of 0.160 μH per conductor. The breakdown strength of air under the waveshape produced by an inductive voltage drop may be taken as 890 kVm^{-1}. These figures when worked out yield an expression for the necessary clearance given by:

$$D = 0.3R + \frac{H}{15n}$$

where, D = clearance in metres
R = resistance to earth in ohms
H = height of building in metres
n = number of down conductors.

It is, in fact, difficult to provide the necessary clearance and it is usually practicable only in small dwellings. In most cases, the alternative of bonding must be adopted.

Metal on the outer surface of a building which is nearer the lightning protective system than the distance given by the above formula should also be bonded to the system. A long piece of metal roughly parallel to the down conductor (e.g. a gutter or a rainwater pipe) should be bonded at each end. Metal services entering the building should be bonded as directly as possible to the earth termination. Large masses of metal, such as a bell frame in a church tower, should be bonded to the nearest down conductor as directly as possible. Short isolated pieces of metal like window frames may be ignored and do not have to be bonded. Similarly, metal reinforcement in a structure which cannot easily be bonded and which cannot itself form part of a down conductor can also be ignored. The danger from such metal is best minimized by keeping it entirely separate from the lightning protection system.

It is perfectly in order for metal cladding or curtain walling which has a continuous conducting path in all directions to be used as part of a lightning protection system. In the extreme case, a structure which is itself a complete metal frame, such as a steel chimney, needs no lightning conductor other than itself. It is enough to earth it effectively.

A structure having reinforcement or cladding forming a close metal mesh in the form of internal reinforcement or screen approaches the conditions of a Faraday Cage, in which any internal metal assumes the same potential as the cage itself. The risk of side flashing is thereby reduced and the recommendations for bonding need not be so strictly adhered to.

The metal bars of concrete reinforcement are tied together by binding wire. Both the bars and the binding wire are usually rusty, so that one does not expect a good electrical contact. Nevertheless, because there are so many of these joints in parallel the total resistance to earth is very low, and experience has shown that it is quite safe to use the reinforcement as a down conductor. Naturally the resistance from air termination to earth must be checked after the structure is complete and if it is too high a separate down conductor must after all be installed.

A building containing explosive or highly flammable materials may need more thorough protection. An air termination network should be suspended above the building or area to be protected, and the conductors should be spaced so that each protects a space formed by a cone having an apex angle of $30°$, i.e. a smaller zone than is adopted for less hazardous buildings. The height of the network should be such that there is no risk of flashover from the network to the building, and the down conductors and earth terminations should be well away from the building. All the earth terminations should be interconnected by a ring conductor buried in the ground. All major metal inside or on the surface of the building should be effectively bonded to the lightning protection system.

It may be difficult to put a radio or television aerial on a roof so that it is within the space protected by the air termination network, and this may present something of a problem. If the down lead is concentric or twin screened, protection can be obtained by connecting the metallic sheath of the cable to the lightning conductor. With a single or twin down lead it is necessary to insert a discharge device between the conductors and an earth lead. In either case metal masts, crossarms and parasitic elements should be bonded to the lightning conductor.

Protection against lightning is dealt with in British Standard Code of Practice BS CP 326.

15 Emergency Supplies

Introduction
There are rare occasions when the public electricity supply fails and a building is left without electricity. In some buildings, the risk of being totally without electricity cannot be taken, and some provision must be made for an alternative supply to be used in an emergency. What form this provision should take is an economic matter which depends on the magnitude of the risk of failure and the seriousness of the consequences of failure. In this chapter, we shall say something about the available methods of providing an alternative supply.

Standby service cable
The Electricity Supply Authority can be asked to bring two separate service cables into the building. They will normally make a charge for this, but it provides security against a fault in one of the cables. It does not, of course, give security against a failure of the public supply altogether.

In heavily built up areas, such as London and other large cities, the public distribution system is in the form of a network and each distribution cable in the streets is fed from a sub-station at each end. The supply system itself thus contains its own standby provision. The only addition the building developer can make is to duplicate the short length of cable from the distributor in the road into the building, and it may be doubted whether the risk of this cable failing is sufficiently great to justify the cost of duplicating it. In rural areas the service cable to individual buildings may be quite long, and may take the form of an overhead line rather than an underground cable. The risk of damage is thus greater than in urban areas and there is much more reason for installing a duplicate cable.

Battery systems
Central battery and individual battery systems have been discussed in Chapter 7 as means of providing emergency lighting. A central battery system can also

provide d.c. power. The next possibility is for the battery to feed a thyristor invertor which then gives a.c. power.

It is difficult to install and keep charged a battery large enough to give the quantities of power needed in a whole building. In building services, in practice batteries are used for emergency lighting but seldom for emergency power.

A battery system will give an emergency supply only for as long as the battery charge lasts. It then becomes dependent on a restoration of the mains supply for long enough to recharge the battery. Thus we can see that this system does not protect against long interruptions of the public supply.

Standby generators

A diesel or gas turbine generator set can be installed in a building to provide electricity when the public supply fails. This is a complete form of protection against all possible interruptions of the main supply. The generator can be large enough to supply all the needs of the building and its output can be connected to the ordinary mains immediately after the Supply Authority's meters and it then provides standby facilities for the entire building. It is cheaper, and may be adequate for the risk to be guarded against, to have a smaller generator serving only the more important outlets. In this case, the distribution must be arranged so that these outlets can be switched from main to emergency supply at one point and so that there is no unintentional path from the emergency generator to outlets not meant to be served by it. In effect the building is divided at the main intake into two distribution systems and only one of them is connected to the emergency change over switch. It is also possible to install a completely separate system of wiring from the emergency generator to outlets quite distinct from the normal ones. This may be the simplest thing to do in a small building or when the emergency supply is required to serve only one or two outlets. It has the disadvantage that individual pieces of equipment have to be disconnected from one outlet and reconnected to another. Whilst this may not be acceptable in a hospital it may be quite in order in a large residence or hostel to have one or two emergency power points into which vacuum cleaners and other domestic equipment can be plugged when the main power supply is interrupted.

Buildings in which standby generators have been installed include poultry farms, chemical process plants, hospitals, telephone exchanges, computer rooms and prisons.

An emergency generator can be started either manually or automatically. A manual start is simple, but it involves a delay during which the building is without power. This delay can be avoided by automatic starting, initated by a sensing unit which detects a drop in the mains voltage. Fig. 153 shows the circuit of a typical mains failure control panel.

When the mains fail, relay 1CC/6 is de-energized and opens the main circuit breaker. It also completes the circuit to the operating coil of relay 2CC/6, thus preparing the circuit for shut down when the mains are restored. Relays VS1,

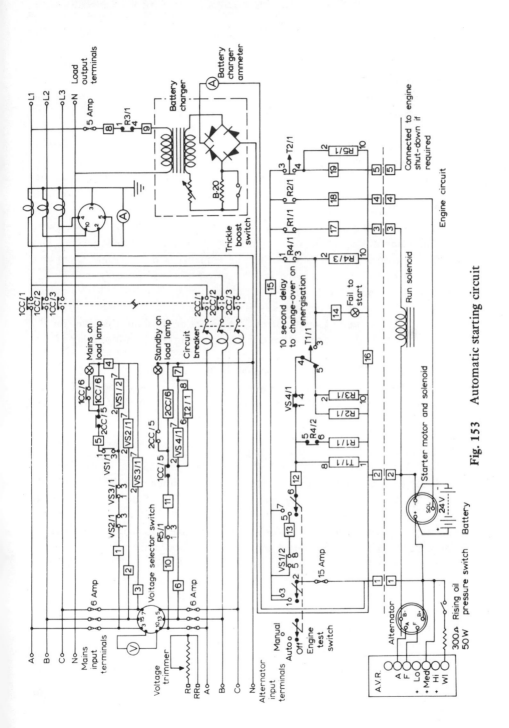

Fig. 153 Automatic starting circuit

VS2 and VS3 are separately operated by each of the three phases, and each has the effect of de-energizing the main relay 1CC/6, so that the system is brought into operation on the failure of any one phase.

When the mains fail, relay VS1/2 is also de-energized and its contacts then bring into operation relays T1, R1, R2 and R3. Relay R1 starts the run solenoid of the diesel engine, relay R2 energizes the starter motor and relay R3 temporarily disconnects the battery charger. If the engine has not started after 10 s, relay T1 de-energizes relays R1, R2 and R3 and lights a fail to start warning lamp. It also energizes relay R4, one of whose contacts breaks the circuit of relay R1 which has the effect of making a restart impossible. Thus, if the engine does not start within 10 s it locks out and nothing further can happen until it receives some manual attention.

If, or perhaps we should say when, the engine starts, relay VS4/1 is energized. This interrupts the operating coil circuits of relays R2 and R3 and prevents relay T1 from energizing relay R4. The starting sequence is thus brought to an end, the battery charger is reconnected and the starter motor stopped.

Relay T2 is now energized and in turn energizes relay R5 which completes the circuit to the operating coil of relay 2CC/6. The latter closes the standby generator. At the same time, it lights the indicator to show that the standby generator is on load and puts a break in the operating coil circuit of relay 1CC/6. This ensures that the mains circuit breaker will not close while the standby circuit breaker is closed. The plant is now running on standby.

When the mains are restored relays VS2 and VS3 are energized and in turn energize relay VS1. The circuit to complete the supply to relay 1CC/6 is thus prepared. Also the circuits to relays T1 and R1 are broken. Relay T1 then breaks the circuit to relay R4 and energizes relays R2 and R3. The starter circuit is kept open by relay R1 which is kept de-energized by relay VS1.

With R1 de-energized the run solenoid is de-energized and the standby set stops. As soon as it shuts down relays VS4 and T2 open. The latter breaks the circuit to relay R5 and the normally open contact of R5 breaks the circuit to relay 2CC/6. This opens the standby circuit breaker and simultaneously completes the circuit to relay 1CC/6, which then closes the main circuit breaker. The standby set is now shut down and the load is back on the mains.

It takes 8 to 10 s for a diesel generator to come to full speed. With the system just described this period is needed to bring the emergency supply into action after the mains have failed and, therefore, during this period there is no supply to the load. In some applications an interruption even of this short duration is not acceptable, and a more complex arrangement is necessary. In one system the diesel engine is coupled to a clutch the other side of which is connected to a squirrel cage induction motor. The induction motor drives an alternator through a flywheel, and the alternator supplies the load. Under normal conditions the induction motor is connected to the mains and the set operates as a motor alternator supplied from the mains. When the mains fail the motor is disconnected from the mains and the diesel engine is started. As soon as it reaches its running speed, the clutch operates and the alternator is driven through the shaft of the motor by the diesel engine. During the time it takes for the engine to come up to speed the alternator is kept going by the flywheel. The automatic controls required for this arrangement are similar to those already described.

Clearly this scheme is much more expensive and involves some permanent losses in the motor alternator set. It is used only for comparatively small power outputs for special purposes, such as telecommunications and power for aircraft landing systems.

Standby generators are normally supplied as complete units on a stand. Fig. 154 is a picture of a typical set. The diesel engine is a normal engine with a governor, and it would be outside the scope of this book to enter on a description of diesel engines. The alternator is directly coupled to the engine and has an automatic voltage regulator. The commonest type of alternator used is the screen protected brushless machine. It is directly coupled to the engine and in smaller sizes may be overhung. In larger sizes it is supported at both ends from the set base plate. A separate exciter is mounted within the casing on the main shaft. In most modern sets the automatic voltage regulator is one of the static

Fig. 154   Standby generator (*Courtesy of* Dale Electric of Great Britain Ltd.).

types. Finally, there is a control panel with voltmeters, ammeters, battery charger, incoming and outgoing terminals and the relays and circuits for the automatic start and stop control. A fuel tank is needed for the diesel engine, but this is normally supplied as a separate item and fixed independently of the generator set, with a short fuel pipe between them.

Diesel engines are noisy and it is prudent to arrange some form of sound-attenuating enclosure. The enclosure must have openings for fresh air to the engine and for the engine exhaust, and these openings will be found to limit the degree of silencing that can be achieved. Several manufacturers supply diesel generator sets complete in an enclosure which provides silencing and is also weatherproof, so that the set can be installed outdoors.

Similar generating sets can, of course, be used to supply power to a building under normal conditions. In the United Kingdom it is not economic for a consumer to generate his own power, but there are still parts of the world where it is a reasonable proposition.

There are, however, cases in industrial countries where it is economic for a consumer to use his own generating plant to supply a peak load. Many factories are supplied on a tariff which includes a charge for the peak instantaneous load. A factory may have a process which takes a fairly steady load during most of the day with a high peak for one or two hours. It may then be economic to limit the power taken from the public supply to rather less than the maximum needed and to make up the deficiency at peak times with the factory's own plant. Fig. 155 shows a load diagram to illustrate this. The public mains are used at all times and are used by themselves so long as the load is less than 300 kVA. As

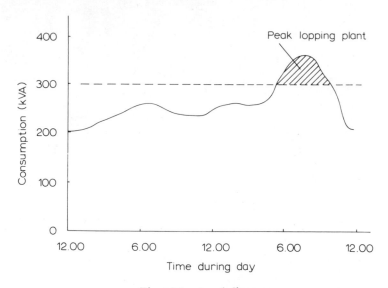

Fig. 155   Load diagram

soon as the load exceeds this figure the factory's own generating plant is started and is run in parallel with the public mains. The power taken from the public mains is limited to a maximum of 300 kVA at all times.

The same type of diesel generator is used for this application as for standby purposes. The start up and shut down sequences are initiated automatically by a kVA meter instead of by a voltage detector, but are otherwise similar to those already described.

Standards relevant to this chapter are:

BS 5266   Code of Practice for the emergency lighting of premises

16 Lifts, Escalators and Paternosters

Introduction

The general design of lifts is very well established, and in this country at least, nearly all the reputable lift manufacturers will design and supply a satisfactory lift as a matter of routine if given the details and the size of building. Nevertheless, the designer of the building electrical services must be able to advise the client about the lifts, to negotiate with the lift suppliers and to compare competing tenders. He must, therefore, know something about the technical details of lifts and we shall accordingly devote this chapter to a brief outline of the subject.

Firstly, we can note that there are three categories of lifts. Passenger lifts are designed primarily for passenger use; goods lifts are mainly for goods but can on occasion carry passengers; and service lifts are for goods only and are of such a size that passengers cannot get into the car. Lift speeds are determined by the number of floors served and the quality of service required. They vary from 0.5 ms^{-1} to 3.0 ms^{-1} in high office blocks.

In deciding the size of car one can allow 0.2 m^2 for each passenger, and when determining the load the average weight of a passenger can be taken as 75 kg. It must, however, be remembered that in many buildings the lift will be used for moving in furniture and the car must be big enough for the bulkiest piece of furniture likely to be needed. The author has made measurements of domestic furniture and has concluded that the most awkward item to manoeuvre is a double bed, which can be up to 1670 mm wide by 1900 long and 360 high. In flats it is unfortunately also necessary to make sure that stretchers and coffins can be carried in the lift. To accommodate these, a depth of 2.5 m is required. The whole car can be made this depth or it can be shallower but have a collapsible extension which can be opened out at the back when the need arises. The lift well must, of course, be deep enough to allow the extension to be opened. In hospitals some of the lifts must take stretchers on trolleys and also hospital beds and these lifts must be the full depth of a complete bed.

Grade of service

The quality of service is a measure of the speed with which passengers can be taken to their destination. It is the sum of the time which the average passenger has to wait for a lift and of the travelling time once he is in the lift. The maximum time a person may have to wait is called the Waiting Time (W.I.) and is the interval between the arrival of successive cars. It depends on the Round Trip Time (R.T.T.) of each lift and on the number of lifts.

The average time a person has to wait is W.I./2. The average time he is travelling is R.T.T./4. The sum of these, W.I./2 + R.T.T./4, is called the grade of service. If there are N lifts, then W.I. = R.T.T./N, and grade of service becomes W.I.$(2 + N)$/4.

It is usual to classify the grade of service as excellent if W.I.$(2 + N)$/4 is less than 45 s, good if it is between 45 and 55 s, fair if it is between 55 and 65 s and casual if it is more than 65 s.

The use of a building will often enable a designer to estimate the probable number of stops during each trip. If this is difficult, then a formula can be developed by probability theory, and is:

$$S_n = n - \left[\left(\frac{P - P_a}{P} \right)^N + \left(\frac{P - P_b}{P} \right)^N + \cdots \left(\frac{P - P_n}{P} \right)^N \right]$$

where, S_n = probable number of stops
n = number of floors served above ground floor
N = number of passengers entering lift at ground floor on each trip
P = total population on all floors
$P_a, P_b, \ldots P_n$ = population on 1st, 2nd $\ldots$ nth floor.

Three to four seconds must be allowed for opening and closing the doors at each stop. A further 1 to 1½ s have to be allowed for each passenger to enter the lift and 1 to 2 s for each passenger to leave.

The travelling time is made up of periods of acceleration, constant speed and retardation. Fig. 156 gives the time versus distance curves for the acceleration normally associated with various lift speeds. On each curve, the point marked X indicates the end of acceleration and start of constant velocity. The retardation is generally taken to be equal in magnitude to the acceleration. Providing the distance between stops is long enough for the lift to reach steady speed before starting to slow, the total travelling time of a round trip is given by:

$$t = \frac{2}{V} (d S_n + D + d)$$

where, t = total travelling time
d = distance during which acceleration takes place
D = distance between ground and top floors
S_n = number of stops between ground and top floors
V = lift speed.

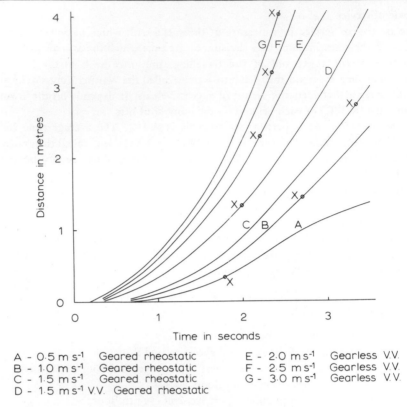

A - 0·5 m s⁻¹   Geared rheostatic   E - 2·0 m s⁻¹   Gearless V.V.
B - 1·0 m s⁻¹   Geared rheostatic   F - 2·5 m s⁻¹   Gearless V.V.
C - 1·5 m s⁻¹   Geared rheostatic   G - 3·0 m s⁻¹   Gearless V.V.
D - 1·5 m s⁻¹ V.V. Geared rheostatic

Fig. 156   Acceleration curves for lifts

The best way of showing how all this data is used to assess the grade of service is by means of an example. Let us assume we are dealing with an office block with eight floors. The heaviest traffic will occur in the morning when people are arriving at work, and we shall assume that we know enough about the occupancy of the building to have been able to estimate that 75% of the work force will arrive in one particular half hour. For estimating the probable number of stops traffic to the first floor can be ignored and we can set out the number of people requiring service as follows:

Floor	2	3	4	5	6	7	8	Total
No. of persons requiring service	36	93	160	85	120	105	63	662

The figures in the second line are 75% of the floor populations, which we assume we have either been given or can guess. The distance between the ground and eighth floor is 25 m.

Table 7 Lift service comparison

	A	B	C	D
Load persons	10	20	10	15
Speed m s^{-1}	1.5	1.5	2.0	2.0
Probable no. of stops per trip (S_n)	5.23	6.35	5.23	5.99
Accelerating distance (d) m	2.60	2.60	2.20	2.20
$d \times S_n$	13.60	16.50	11.50	13.20
Distance between ground and top floors (D) m	25.00	25.00	25.00	25.00
$(dS_n + D + d)$	41.20	44.10	38.70	40.40
$\dfrac{2(dS_n + D + d)}{2 \times \text{speed}}$ = travelling time (s)	55.0	59.0	38.00	40.00
Door opening time (s)	21.00	28.00	21.00	24.00
Passengers entering and leaving (s)	25.00	50.00	25.00	37.00
Total travelling time	101	137	84	101
10% margin	10	13	8	10
R. T. T. (s)	111	150	92	111
No. of lifts	4	3	4	3
No. of trips per lift in 30 min	16	12	19	16
No. of persons per lift in 30 min	160	240	190	240
Total no. of persons carried in 30 min	640	720	760	720
W.I (s)	28	50	23	37
$\dfrac{\text{W.I}}{4}(2 + N)$	42	62	35	46
Grade of service	Excellent	Fair	Excellent	Good

Calculation of S_n $S_n = n - \displaystyle\sum_{i=2}^{i=8}\left(\dfrac{P - P_i}{P}\right)^N$ $P = 662$
$n = 7$

i	P_i	$P - P_i$	$\dfrac{P - P_i}{P}$	$\left(\dfrac{P - P_i}{P}\right)^{10}$	$\left(\dfrac{P - P_i}{P}\right)^{15}$	$\left(\dfrac{P - P_i}{P}\right)^{20}$
2	36	626	0.95	0.60	0.46	0.36
3	93	569	0.86	0.22	0.10	0.05
4	160	502	0.76	0.06	0.01	0.00
5	85	577	0.858	0.22	0.10	0.05
6	120	542	0.82	0.12	0.04	0.02
7	105	557	0.84	0.18	0.08	0.03
8	63	599	0.905	0.37	0.22	0.14
			Σ 1.77		1.01	0.65
			$S_n = 7 - \Sigma$ 5.23		5.99	6.35

The round trip time can be calculated and hence it is possible to calculate the number of lifts needed to carry 662 people in 30 min. From this, the grade of service can be obtained. If the calculation is set out in tabular form, different combinations can be easily compared. This has been done in Table 7. A 10% margin is added to the calculated total time to allow for irregularities in the time interval between different lifts in the bank of lifts.

It can be seen that in this example the most satisfactory arrangement is four lifts each taking 10 persons at a speed of 1.5 ms^{-1}. 2.0 ms^{-1} would be unnecessarily extravagant.

It will be found that where the service is not so concentrated lower speeds are sufficient. For this reason, it should not be necessary to use speeds of more than 0.75 or 1.0 ms^{-1} in blocks of flats.

Accommodation

The machine room for the lifting gear is normally at the top of the lift shaft or well. It can be at the bottom or even beside the well, and in the latter case it can

Table 8   Lift dimensions

General purpose passenger lifts

Load persons	Speed $(m\ s^{-1})$	Well		Machine room			Top landing to M/C room floor (m)	Pit depth (m)
		Width (m)	Depth (m)	Width (m)	Length (m)	Height (m)		
8	1.0	1.80	1.90	3.10	4.80	2.60	4.00	1.70
10	0.75	2.00	1.90	3.10	5.00	2.60	4.00	1.60
10	1.0	2.00	1.90	3.10	5.00	2.60	4.00	1.70
10	1.5	2.00	1.90	3.10	5.00	2.60	4.20	1.70
16	0.75	2.60	2.20	3.50	5.30	2.70	4.10	1.70
16	1.00	2.60	2.20	3.50	5.30	2.70	4.20	1.90
16	1.50	2.60	2.20	3.50	5.30	2.70	4.30	1.90
20	0.75	2.60	2.50	3.50	5.60	2.70	4.10	1.70
20	1.00	2.60	2.50	3.50	5.60	2.70	4.20	1.90
20	1.50	2.60	2.50	3.50	5.60	2.70	4.30	1.90

High speed passenger lifts

Load persons	Speed $(m\ s^{-1})$	Well		Machine room			Top landing to M/C room floor (m)	Pit depth (m)
		Width (m)	Depth (m)	Width (m)	Length (m)	Height (m)		
12	2.5	2.20	2.20	3.20	7.50	2.70	6.80	2.80
16	2.5	2.60	2.30	3.20	8.00	2.70	6.80	2.80
16	3.5	2.60	2.30	3.20	8.00	3.50	6.90	3.40
20	2.5	2.60	2.60	3.20	8.30	3.50	6.20	2.80
20	3.5	2.60	2.60	3.20	8.30	3.50	7.10	3.40
20	5.0	2.60	2.60	3.20	8.30	3.50	8.20	5.10

Table 8 Continued

General purpose goods lifts

Load (kg)	Speed ($m\,s^{-1}$)	Well		Machine room			Top landing to M/C room floor (m)	Pit depth (m)
		Width (m)	Depth (m)	Width (m)	Length (m)	Height (m)		
500	0.5	1.80	1.50	2.00	3.70	2.40	3.80	1.40
1000	0.25	2.10	2.10	2.10	4.30	2.40	3.80	1.50
1000	0.50	2.10	2.10	2.10	4.30	2.40	3.80	1.50
1000	0.75	2.10	2.10	2.10	4.30	2.40	3.80	1.50
1500	0.25	2.50	2.30	2.50	4.50	2.70	4.00	1.50
1500	0.75	2.50	2.30	2.50	4.50	2.70	4.20	1.80
1500	1.00	2.50	2.30	2.50	4.50	2.70	4.20	1.80
2000	0.25	2.80	2.40	2.80	4.70	2.90	4.10	1.50
2000	0.75	2.80	2.40	2.80	4.70	2.90	4.50	1.80
2000	1.00	2.80	2.40	2.80	4.70	2.90	4.50	1.80
3000	0.25	3.50	2.70	3.50	5.00	2.90	4.20	1.50
3000	0.50	3.50	2.70	3.50	5.00	2.90	4.40	1.70
3000	0.75	3.50	2.70	3.50	5.00	2.90	4.50	1.80

Heavy duty goods lift

Load (kg)	Speed ($m\,s^{-1}$)	Well		Machine room			Top landing to M/C room floor (m)	Pit depth (m)
		Width (m)	Depth (m)	Width (m)	Length (m)	Height (m)		
1500	0.50	2.60	2.40	2.60	4.80	2.70	4.80	1.70
1500	0.75	2.60	2.40	2.60	4.80	2.70	4.80	1.80
1500	1.00	2.60	2.40	2.60	4.80	2.70	4.80	1.80
2000	0.50	2.90	2.50	2.90	5.00	2.90	4.80	1.70
2000	0.75	2.90	2.50	2.90	5.00	2.90	4.80	1.80
2000	1.00	2.90	2.50	2.90	5.00	2.90	4.80	1.80
3000	0.50	3.50	2.80	3.50	5.30	2.90	4.80	1.70
3000	0.75	3.50	2.80	3.50	5.30	2.90	4.80	1.80
3000	1.00	3.50	2.80	3.50	5.30	2.90	4.80	1.80
4000	0.50	3.50	3.40	4.00	6.20	2.90	5.20	1.70
4000	0.75	3.50	3.40	4.00	6.20	2.90	5.20	1.80
5000	0.50	3.60	4.00	4.00	6.80	2.90	5.20	1.70
5000	0.75	3.60	4.00	4.00	6.80	2.90	5.20	1.80

be at any height, but from these positions the ropes must pass over more pulleys so that the overall arrangement becomes more complicated. It is, therefore, better to provide space for the machine room at the top of the well. Room must also be left for buffers and for inspection at the bottom, or pit, of the well. The sizes of the machine and pit rooms must ultimately be agreed with the lift manufacturers, but for preliminary planning before a manufacturer has submitted a quotation the dimensions in Table 8 may be taken as a guide.

Drive

Nearly all lifts use a traction drive. In this, the ropes pass from the lift car round a cast iron or steel grooved sheave and then to the counterweight. The sheave is secured to a steel shaft which is turned by the driving motor. The drive from the motor to the shaft is usually through a worm gear. The force needed to raise or lower the lift car is provided by the friction between the ropes and the sheave grooves. The main advantage of the traction drive is that if either the car or counterweight comes into contact with the buffers the drive ceases and there is no danger of the car being wound into the overhead structure. Other advantages are cheapness and simplicity.

Fig. 157 shows a traction machine which has a squirrel cage motor permanently coupled to the gear by a vee rope drive. There is a brake working on a brake disc on the end of the worm shaft and the grooved sheave is easily discernible at the top of the machine.

The only other kind of drive is the drum drive. In this case, one end of the car ropes and one end of the counterweight ropes are securely fastened by clamps on the inside of a cast iron or steel drum, the other ends being fastened to the car and counterweight respectively. One set of ropes is wrapped clockwise round the drum and the other set anti-clockwise, so that one set is winding up as the other set is unwinding from the drum. As the car travels, the ropes move along

Fig. 157   Traction machine (*Courtesy of* The Express Lift Co. Ltd.)

the drum in spiral grooves on its periphery. The drum drive suffers from the disadvantage that as the height of travel increases the drum becomes large and unwieldy. It has been almost entirely superseded by the traction drive.

Fig. 158 shows various arrangements of ropes for traction drives. Fig. 158b shows a double wrap drive, in which each rope passes over the sheave twice. The increased length of contact between rope and sheave increases the maximum available lifting force, or alternatively permits a lower coefficient of friction for the same lifting force. Fig. 158c shows a double wrap two to one system in

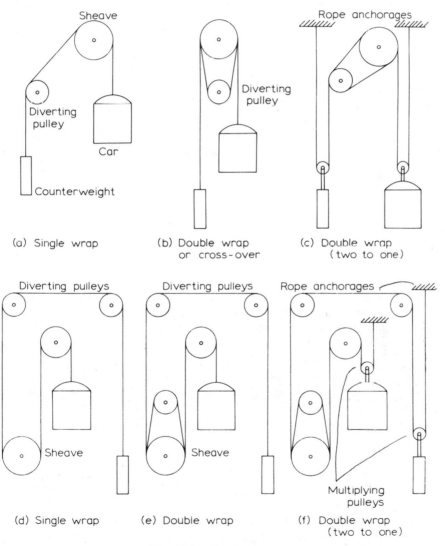

Fig. 158 Roping systems

which the speed of the car is half the peripheral speed of the sheave. Figs. 158d, e and f correspond to Figs. 158a, b and c but with the winding machine at the bottom of the well.

Compensating ropes are sometimes fitted on long travel lifts in order to make the load on the motor constant by eliminating the effect of the weight of the ropes. A simple method of doing this is shown in Fig. 159.

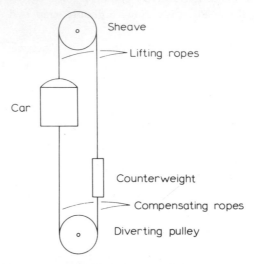

Fig. 159   Compensating ropes

Motors

A lift motor should have a starting torque equal to at least twice the full load torque; it should be quiet and it should have a low kinetic energy. The last requirement is necessary for rapid acceleration and deceleration and also for low wear in the brakes. The theoretical power needed can be calculated from the lifting speed and the greatest difference between the weights of car plus load and counterweight. The actual power will depend on the mechanical efficiency of the drive which can be anything from 30% to 60%. Suitable motor sizes for various lifts are given in Table 9. The acceleration is settled by the torque-speed characteristic of the motor and the ratio of motor speed to lift speed.

Types of motor

In most cases, in the United Kingdom at least, a three phase a.c. supply is available in a building which is to have a lift installation. For lift speeds up to about 0.5 ms^{-1} a single speed squirrel cage motor is suitable, although it has a high starting current and tends to overheat on duties requiring more than 100 starts an hour. Better performance is obtained with a wound rotor motor which is accelerated by the removal of rotor resistances in several steps.

Table 9   Approximate lift motor ratings (motor ratings in kW)

Low efficiency geared lifts

Car speed (m s⁻¹)	Contract load (kg)					
	250	500	750	1000	1500	2000
0.25	2	3.5	5	5	8	10
0.50	3	5.0	8	10	15	20
0.75	4	8.0	10	15	20	25
1.00	5	10	12	18	25	35
1.25	6	12	15	20	30	45
1.50	7	15	20	25	40	50

High efficiency geared goods lifts

Car speed (m s⁻¹)	Contract load (kg)						
	250	500	750	1000	1500	2000	3000
0.25	1	2.0	3	4.0	6.0	8	10
0.50	3	4.0	6	7.5	10.0	14	20
0.75		5.0	8	10.0	12.5	20	30
1.00		7.5	10	14.0	20.0	25	

High efficiency geared passenger lifts

Car speed (m s⁻¹)	No. of passengers					
	4	6	8	10	15	20
0.50	3	3	4.0	5.0		
0.75		5	7.5	7.5	12	15
1.0			10	10	15	18
1.5				12	17.5	22

For speeds between about 0.5 ms⁻¹ and 1.25 ms⁻¹, it becomes necessary to use a two speed motor in order to have a low landing speed. The squirrel cage motor can be wound to give two combinations of poles, thus giving two speeds. The motor of the machine illustrated in Fig. 157 is of this kind. Alternatively, two separate windings can be put in the same slots but this is more expensive and the motor is harder to repair if a winding is damaged.

Speed changes can be obtained with slip ring motors, but this usually requires the use of two separate rotor windings. The rotor connections must be changed at the same time as the stator connections, which complicates the control circuit. This method does, however, give better performance.

A number of lifts have in the past made use of a tandem motor. This consisted of one wound rotor and one squirrel cage rotor mounted in tandem on the same shaft. The two frames were bolted together to make a single two-bearing unit. These motors gave a higher efficiency than a two-speed squirrel cage motor and could be used for heavier duties, but they cost nearly twice as much. The price disadvantage and the ease with which a.c. can now be rectified to supply d.c. Ward Leonard systems have so reduced the demand that tandem motors are no longer made, but the engineer concerned with existing installations will still need to know of their existence.

A.c. commutator motors can also give the desired speed variation, but they are expensive and noisy and are very seldom used for lifts.

For speeds above 1.5 ms^{-1} the same types of motor could be used as for the lower speeds, but in fact they very rarely are. At the higher speeds, it is possible to design a d.c. motor to run at a speed which makes reduction gearing to the drive unnecessary. The a.c. supply is therefore used to drive a variable voltage motor generator set which supplies the d.c. to the lift driving motor.

If only a single phase supply is available, a repulsion-induction motor can be used for speeds up to 0.5 ms^{-1}. For higher speeds it is better to use the a.c. to drive a motor generator and have a d.c. lift machine.

When d.c. is used for speeds less than 0.5 ms^{-1}, a single speed shunt or compound wound motor is employed. When the lift is decelerating the machine runs as a generator, and in the case of a compound wound motor this makes special arrangements necessary. For speeds between 0.5 ms^{-1} and 1.25 ms^{-1} two-speed shunt motors are used, so that a lower speed is available for good levelling at the landings. The increase from low to high speed is obtained by the insertion of resistance in the shunt field. For speeds of 1.5 ms^{-1} and above, a d.c. shunt wound motor running between 50 and 120 r.p.m. is employed. At these speeds the motor can be coupled directly to the driving sheave without any gearing. Because of the size of the motor it is not possible to vary the speed by more than 1.5 to 1 by field control, and speed is usually controlled by the Ward Leonard method. The absence of gearing increases the overall efficiency, improves acceleration and results in smoother travelling. A picture of a gearless machine is shown in Fig. 160.

The Ward Leonard method is also used for geared d.c. machines because it gives smoother acceleration and deceleration, can regenerate to the supply mains and simplifies the controller. The contactors in the controller need handle only small currents instead of full power as in the case of rheostatic controllers.

Variable voltage d.c. can also be provided by a grid controlled mercury arc rectifier.

Brakes

Lift brakes are usually electromagnetic. In the majority of cases, they are placed between the motor and the gearbox; in a gearless machine the brake is keyed to the sheave. The shoes are operated by springs and released by an electromagnet

Fig. 160 Gearless lift machine (*Courtesy of* The Express Lift Co. Ltd.)

the armature of which acts either directly or through a system of links. A typical brake is shown in Fig. 161.

Lift cars

Passenger cars should be at least 2.00 m high, and preferably 2.15 m or more. They can be made to almost any specification, but most manufacturers have certain standard finishes which the client should choose from.

Lift cars consist of two separate units, namely the sling and the car proper. The sling is constructed of steel angles or channels and the car is held within the sling. The sling also carries the guide shoes and the safety gear. The car is sometimes insulated from the sling frame by anti-vibration mountings. Goods cars are of rougher construction than passenger cars but otherwise follow the same principles.

All electrical connections to a car are made through a multi-core hanging flexible cable. One end of this is connected to a terminal box under the car, and the other end to a terminal box on the wall of the well approximately half way down.

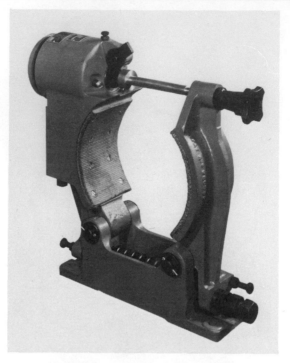

Fig. 161   Lift brake (*Courtesy of* Dewhurst & Partner Ltd.)

Counterweights

A counterweight is provided to balance the load being carried. As the load carried varies, the counterweight cannot always balance it exactly; it is usual for the counterweight to balance the weight of the car plus 50% of the maximum load to be taken in the car. A typical counterweight is shown in Fig. 162. It consists of cast iron sections held in a steel framework and rigidly bolted together by tie rods. The lifting ropes are attached to eye bolts which pass through the top piece of the frame.

Guides

Both the car and the counterweight must be guided in the well so that they do not swing about as they travel up and down. Continuous vertical guides are provided for this purpose. They are most commonly made of steel tees, and there are standard tees made especially for use as lift guides. The guides are fastened to steel plates by iron clamps at intervals of about 2 m and these plates are secured to the sides of the well. They may be secured by bolts passing through the wall of the well and held by back plates on the other side or by being attached to angle irons or channels which are in turn built into the wall. The latter is the usual practice with concrete building construction.

Fig. 162   Counterweight (*Courtesy of* The Express Lift Co. Ltd.)

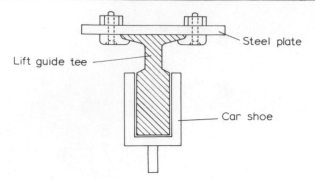

Fig. 163   Shoe on lift guide

Guide shoes are fitted on the car and on the counterweight and run smoothly on the guides. Fig. 163 shows a shoe on a guide. For smooth running, the guides must be lubricated and various types of automatic lubricators have been designed for lift guides. The commonest kind make use of travelling lubricators mounted on the car and counterweight. More recently unlubricated guides have been used with shoes lined with carbon or PTFE.

Doors

Solid doors have now entirely superseded collapsible mesh gates. They are quieter, stronger and safer. It is usual now for the car and landing doors to be operated together. If the entrance to the car is not to be much narrower than the car itself then in the open position the door must overlap the car. To accommodate this, the well must be wider than the car. This will be clear from the plan of a typical lift installation shown in Fig. 164.

Doors can be opened and closed manually, but it is more usual to have them power operated. In order not to injure passengers caught by closing doors, the drive has to be arranged to slip or reverse if the doors meet an obstruction. Every lift car door must have an interlock which cuts off the supply to the lift controller when the door is open. This can be a contactor which is pushed closed by the door and falls open by gravity or spring action when the door opens.

The landing door must be locked so that it cannot be opened unless the car is in line with the landing. The most usual way of doing this is by means of a lock which combines a mechanical lock and an electrical interlock. The electrical interlock ensures that there is no supply to the controller unless the gate is locked. The mechanical part can be unlocked only when a cam on the car presses a roller arm on the lock; thus the landing door can only be opened when the car is at the landing. The controls withdraw the cam when the car is in motion and return it only as the car approaches a floor at which it is to stop. This makes it impossible for anyone to open a landing door as the car passes the landing if the car is not stopping there.

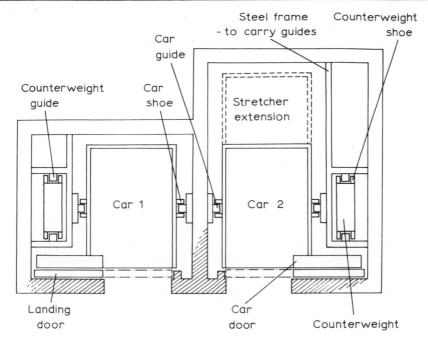

Fig. 164   Plan of typical lift arrangement

Indicators

Indicators are available for showing when the car is in motion, the direction of travel and the position of the car in the well. A position indicator is usually installed in the car, and in many cases also at each landing. It is cheaper to have only a direction indicator at the landings, and a common arrangement is to have a position indicator in the car and at the ground floor with direction indicators at the other landings.

The wiring diagram of a landing indicator is given in Fig. 165. Switches U1 − 4 close during the upward motion of the car, whilst DG and D1 − 3 close when the car is at half the distance between floors. The switches are actually on a floor selector machine in the machine room. A picture of such a machine is shown in Fig. 166. It contains a shaft which is driven by a rope attached to the car. The angular position of the shaft is, therefore, a measure of the vertical position of the car in the well. There are a number of arms or cams on the shaft, and as the shaft rotates each of these in turn operates a switch. Each arm is set to operate its switch at a particular position of the car.

Safety devices

Every lift car must have a safety gear which will stop it if its speed increases above a safe level. The motor and brake circuits should be opened at the same

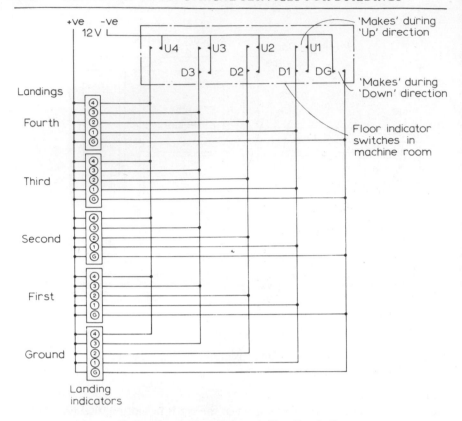

Fig. 165   Wiring diagram of landing indicator

time as the safety gear operates. If the lift travel is more than about 9 m the safety gear should be operated by an overspeed governor in the machine room.

For speeds up to 0.8 ms^{-1} instantaneous gear is generally used. This consists of a pair of cams just clear of each guide, one on each side of the guide. The cams have serrated edges and are held away from the guide by springs. A safety flyrope passes from the safety gear over a top idler wheel to the counterweight. Tension on the safety rope causes the cams to come into instantaneous contact with the guides and they then clamp the car to the guides. The safety rope comes into tension if the lifting ropes break. Alternatively, it can be connected to an overspeed governor. One type of governor is shown in Fig. 167. It has a pulley driven from the car by a steel rope. Flyweights are mounted on the pulley and linked together to ensure that they move simultaneously and equally. The flyweights move against a spring which can be adjusted to give the required tripping speed. As the speed increases, the weights move out against the spring force and at the tripping speed they cause a jaw to grip the rope, which produces the tension necessary to operate the cams.

Fig. 166 Floor selector machine (*Courtesy of Dewhurst & Partners Ltd.*)

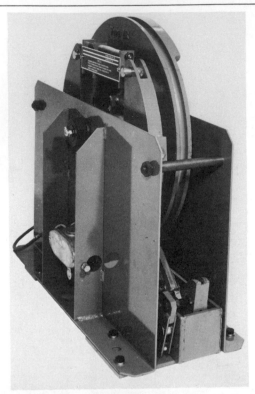

Fig. 167   Lift governor (*Courtesy of* Dewhurst & Partners Ltd)

For lifts at higher speeds, gradual wedge clamp safety gear is used. This also works by clamping the car to the guides, but the clamps are forced against the guides gradually and so bring the car to rest more smoothly. The clamps can be brought into play by screw motion or by a spring.

Another type of safety gear used on high speed lifts is the flexible guide clamp, an example of which is illustrated in Fig. 168. It consists of two jaw assemblies, one for each guide, mounted on a common channel under the car. The jaw assembly has a pair of jaws with a gib in each jaw. Tension on the governor rope resulting from operation of the governor pulls the operating lever and causes the gibs to move up the jaws. The consequent wedging action of the gibs between the jaws and the guides compresses the jaw spring to produce a gripping force on the guides which gives a constant retardation.

A lift must also have upper and lower terminal switches to stop the car if it overruns either the top or bottom floor. These can take the form of a switch on the car worked by a ramp in the well, or they may consist of a switch in the well worked by a ramp on the car. There should be a normal stopping switch and a fixed stopping switch at each end of the travel.

Clearance must be allowed for the car at the top and bottom of the well to

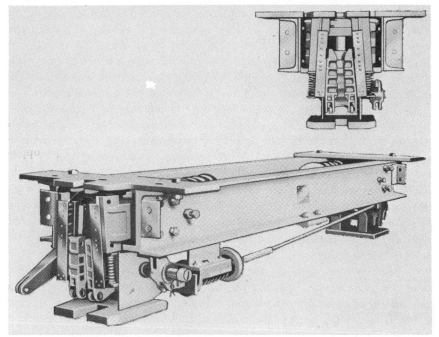

Fig. 168   Flexible guide clamp safety gear (*Courtesy of* The Express Lift Co. Ltd.)

give it room to stop if the normal terminal switch fails and is passed and the terminal switch operates. The clearances can be as given in Table 10. The bottom clearance given in the table includes the buffer compression.

The final safety device consists of buffers in the well under the car and under the counterweight. For low speed lifts they can be made as volute or helical springs, but for high speed lifts oil buffers are used.

Landing

As it stops, the car must be brought to the exact level of the landing. With an automatic lift, this depends on the accuracy with which the slowing and stopping devices cut off the motor current and apply the brake. Levelling is affected by the load being carried; a full load travels further than a light load

Table 10   Lift clearances

Lift speed (m s⁻¹)	Bottom clearance (m)	Top clearance (m)
0–0.5	0.330	0.455
0.5–1.0	0.410	0.610
1.0–1.5	0.510	0.760

when coming to rest from a given speed on the downward trip and less far on the upward trip. To overcome this, it is desirable that the car should travel faster when carrying a full load up than when travelling up empty. A motor with a rising characteristic would be unstable, but the desired effect can be easily achieved with variable voltage control. The rising characteristic is needed only at the levelling speed, which is from about 1/6 to 1/20 of the maximum speed.

Automatic slowing and stopping of a car is often done by means of a floor selector machine, which we have already described and illustrated in Fig. 166. In addition to the switches which energize the indicators, the machine can have other switches at positions appropriate to bring in the slowing and stopping contactor coils. Floor selectors can be constructed in a number of ways, but the function they perform is always the same. To ensure that the car starts in the right direction, direction switches are fitted in the well, one at each landing. They are operated by a ramp on the car, and the effect is that all call switches below the car are connected to the main down contactor while all call switches above the car are connected to the main up contactor.

There are several methods available for final levelling from the position at which the floor selector stops the car. One way is to have a three level ramp at each landing engaging a three position switch on the car. If the car stops below or above the landing, the switch engages with the appropriate level of the ramp, and this results in the car's moving up or down as necessary. The centre level of the ramp cuts off the motor supply. When the car is to pass a landing without stopping the operating arms of the switch on the car are withdrawn so that they do not engage the ramp. The low levelling speed is sometimes provided by an auxiliary motor which drives the lift machine through a vee rope drive and a friction clutch.

The various switches required in the well to actuate the controls can be replaced by inductors. An inductor is a solenoid switch with an air gap in the magnetic path, and it is fitted to the car. At the appropriate level in the well there is a projecting steel plate, so positioned that it passes through the air gap of the inductor. When the switch passes this plate, the latter diverts the magnetic flux and thus operates the armature of the solenoid.

Type of control

Manual control requires only a switch in the car with which the attendant starts and stops the car. Levelling depends entirely on the skill of the attendant. A call button at each landing illuminates an indicator in the car, thus telling the attendant at which floor he is required. In department stores the lift normally stops at every floor on every trip. Call buttons are not, therefore, needed at the landings and stopping and landing are carried out automatically. The attendant's switch is used only to start the lift. For banks of high speed lifts, signal control can be used. With this method, a passenger waiting for a lift presses either an UP or a DOWN button on the landing according to the direction in which he wishes to travel. The first car moving in the desired direction automatically stops for him, and as he enters he tells the attendant which floor he wishes to go to. The attendant then presses a button for that floor and moves the car switch to START. Stopping is done automatically at all selected floors and the car is also automatically stopped at any intermediate floor at which waiting passengers have pressed the appropriate call button.

However, automatic control has almost completely replaced all forms of manual control. Perhaps the only places in which new lifts are likely to be attendant-operated are department stores, where the lift attendant doubles as enquiry service and can tell customers where to find the section they want.

An automatic control system has a single call button at each landing and a button for each floor in the car. A passenger presses the car button for the floor he desires and the lift automatically travels there. Calls made from landings while the car is in motion are ignored. This becomes unsatisfactory if there is more than a very light demand for the lift because if more than one person is waiting, for any one of them to get it depends on his pressing the call button a second, third or fourth time just after it stops and before anyone else does so. The lift service thus becomes a lottery, which most people find infuriating.

The difficulty is overcome by Automatic Collective Control. Each landing has both an UP and a DOWN button, and there is a set of floor buttons in the car. Every button pressed registers a call, and up and down calls are answered during up and down journeys respectively, in the order in which the floors are reached. The order in which the buttons are pressed does not affect the sequence in which the car stops at the various floors, and all calls made are stored in the system until they have been answered. Down calls made while the lift is travelling up are kept until after the up journey is finished, and up calls made while the lift is moving down are similarly kept until that trip is finished.

The system can be modified to work as a collective system in the down direction and as a simple automatic system in the up direction. It is then known as Down Collective. This version is sometimes used in blocks of flats and is based on the assumption that occupants and their visitors travelling up like to go straight to their own floors, but that everyone going down wants to get off at the ground floor. Thus upward travellers should be able to go straight to their own floor without interference, while downward travellers are less likely to be irritated by intermediate stops to pick up other passengers going to the same destination. This reasoning ignores milkmen, postmen and other delivery workers, and the author of this book finds it unconvincing. Nevertheless, it appears to be popular with many authorities.

Duplex control is used when two lifts are installed in adjacent wells. The landing buttons serve both lifts. Landing calls are stored and allotted to the cars one at a time as the cars finish journeys already in progress. For three lifts working together Triplex Control is used.

Finally, we may mention that it is also possible to arrange lifts with dual control so that they can be used either with or without an attendant.

Controllers

The lift motors are started, stopped and reversed by contactors. The operating coils of the contactors are energized at the appropriate times by relays which are connected in a circuit to give the required scheme of operation. The assembly of contactors, relays and associated wiring forms the controller, which is usually placed in the motor room. Sufficient space must be left round the controller for maintenance, and it must also be placed so that a maintenance technician working at it cannot accidentally touch a moving part of the lift machine. A controller for a variable voltage gearless machine is shown in Fig. 169. In this example, the floor selector is included in the controller, and can be seen at the right hand side of the picture.

A controller circuit is necessarily complex because it contains many interconnected relays; but each relay performs only one function and they are arranged to energize each other in a logical sequence to achieve the operation required. The resulting wiring diagram may perhaps be described as complex but not complicated. In a book which deals with electrical services as a whole and devotes only one chapter to lifts, it is not practicable to give a full description of circuits for all the possible modes of control, and we shall do no more than explain the working of a simple automatic system. It will have to serve as a

Fig. 169   Variable voltage motor (*Courtesy of* Dewhurst & Partners Ltd)

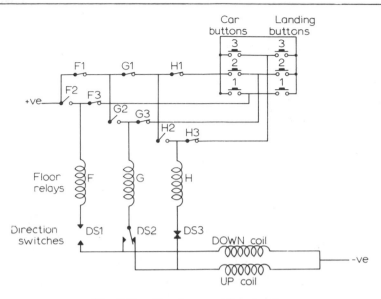

Fig. 170   Simple controller circuit

model illustrating principles which can be extended and adapted for other systems.

The circuit of our simple model is given in Fig. 170. There is a single way direction switch at each terminal floor, and a two way direction switch at each intermediate floor. If the car is travelling upwards the direction switch changes from UP to DOWN as the car passes it, and vice versa if the car is travelling downward. Thus all direction switches above the car have connected their respective floor relays to the coil of the UP contactor. If at this point either a landing button or a car button for a floor above the car is pressed, a circuit is completed through the floor relay and the UP contactor coil. One of the contacts of the floor relay then holds the relay in, so that the circuit remains made after the button is released. The opening of the two normally closed contacts on the relay prevents the completion of a circuit through any other relay, and thus prevents any other button from making an effective call until the one in hand has been dealt with.

The completion of the circuit through the UP contactor coil starts the lift in the up direction. As the car reaches the floor to which it has been commanded, the car ramp operates the direction switch; this breaks the circuit and de-energizes the UP contactor coil. The same action also operates the brake and thus stops the car at the required landing.

Collective controllers and duplex collective controllers have a larger number of relays and a more complex circuit. There seems little point in embarking on a full account which would be long, tedious to prepare and certain to be skipped by the majority of readers, and we shall not attempt to do so. The engineer called upon to investigate the internals of a controller will in any case have to obtain the circuit of that particular machine from its manufacturer, and for the purpose of explaining the principles involved, the circuit of Fig. 170 should be adequate.

The controllers we have described answer landing calls separately for up and down journeys. The landing calls indicate in which direction the passenger wishes to travel, but not his actual destination, and the order in which passengers are picked up and put down is determined by a simple, but rigid, pattern. If the destination of all waiting passengers were known, one could calculate all possible ways of combining the desired journeys and by comparing them arrive at the most economic sequence of trips for the cars. Such a calculation can readily be done by a digital computer, and it has been suggested that a computer could replace a conventional controller for large lift installations. At each landing, there is a complete set of floor buttons and a waiting passenger presses the button for the floor he wishes to go to. The digital computer allocates the call to the most suitable car and an illuminated sign tells the passenger which car will serve him. He waits for this car, enters it and is then taken to his destination without pressing further buttons. Simulation of this system on a computer has shown that it can reduce waiting time sufficiently to enable four cars to give a service requiring six cars with conventional control. The system has been described and strongly advocated in published papers. We have referred to it here because of the likelihood that it will come into use, but as far as the author is aware, it has not yet actually been installed.

Escalators

An electrical services engineer should also know something about escalators. These are moving staircases. They consist essentially of pivoted steps linked to each other and pulled along by an endless chain. The steps have guide pins which move in tracks on either side of the tread arranged so that the steps come out of the concealed section horizontally, are then pulled or pivoted into the shape of a staircase and finally return to the horizontal before returning back into the concealed section. The steps complete their circuit within the concealed section, and on this part of their travel they are flat. This is illustrated in Fig. 171.

The surface of each tread has grooves parallel to the direction of motion. The stationary platforms at the top and bottom of the escalator have fixed combs which mesh with the grooves in the treads in order to ensure a smooth run in and out of the treads under the fixed floor. The stationary platform is actually a floor plate over the recess under the moving staircase and covers the working mechanism at the top and bottom landings. It has an extension known as a combplate and the combplate carries the projecting comb teeth.

An escalator also has a balustrade with a handrail, and the handrail moves on an endless chain in step with the stairs. There are separate chains for the handrail

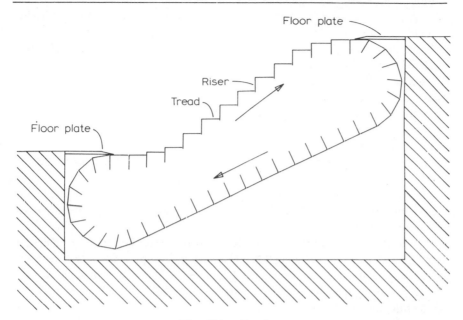

Fig. 171   Escalator

and the steps but they are both driven through a gearbox from the same motor. The usual speed of an escalator is 0.5 ms^{-1} and it ought never to exceed 0.75 ms^{-1}. The inclination varies between $27°$ and $35°$ to the horizontal.

The drive and transmission have to carry the total load on the escalator. Since people do not stand at even and regular intervals on the whole staircase the load averaged over the whole length of the escalator is less than the maximum load on individual treads. The peak load on each tread is of concern to the structural designer but the electrical engineer concerned with the power requirements can use the average passenger load taken over the total area of exposed treads. This average can be taken as 290 kgm^{-2}.

An escalator must have a brake which has to fail safe if there is an interruption to the electrical supply. The brake is therefore applied by a spring or a hydraulic force and is held off against the mechanical force by an electrically energized solenoid. As is the case with lifts, there is also provision for releasing the brake manually and handwinding the escalator.

Since an escalator is in continuous operation there are no passenger controls, but there must be an on/off switch which can be worked by a responsible person in charge of the premises. It would be dangerous for the escalator to be started or stopped by someone who could not see the people on it and the switch must be in a place from which the escalator can be seen. Such a place is almost inevitably in reach of the public using the escalator, who ought not to be able to work the switch, and so the switch must be a key operated one. The British

Standard on Escalators (BS 2655) requires a key operated starting switch to be provided at both ends of the escalator.

Emergency stop switches are provided in the machinery spaces under the escalator and also in positions accessible to the public at the top and bottom of the escalator. The operation of any of these switches disconnects the electrical supply from both the driving machine and the brake. The removal of the supply from the brake allows the mechanical force to apply the brake and bring the escalator to a halt. Some escalators are fitted with a speed governor which similarly disconnects the electrical supply from the drive and brake. There are further safety devices to disconnect the supply if one of the driving chains breaks.

Most escalators are reversible. The driving motor is a squirrel cage induction motor and the drive is reversed by contactors which change the phase sequence of the supply to the motor.

A travellator differs from an escalator in being either horizontal or having a very small slope not exceeding $12°$. This makes it unnecessary for it to form steps and passengers are conveyed on a continuous platform. The upper surface of the platform must have grooves parallel to the direction of motion which mesh with the combplates. In other respects a travellator or moving walkway is designed in exactly the same way as an escalator.

Paternosters

Whilst paternosters are a type of lift they also have similarities with escalators and it is more convenient to discuss them after the latter. A paternoster is a lift which has a series of small cars running continuously in a closed loop. It is difficult to explain this clearly in words but it should be clear from Fig. 172. The cars are open at the front and move slowly enough for people to step in and out of them whilst they are in motion, just as they step on and off an escalator. In fact a paternoster can perhaps be thought of as a vertical escalator. To make it safe for people to get on and off whilst the cars are in motion the speed must be less than 0.3 ms^{-1}.

The cars are constructed in the same way as ordinary lift cars but do not have doors and are not large enough to take more than one person each. In practice this means that the cars are less than 1.0 m x 1.0 m in plan. They must of course be of normal height. The front of the floor of each car is made as a hinged flap. This ensures that if a person has one foot in the car and one on the landing he will not be thrown off balance as the car moves up. Since the cars move in a continuous loop they provide their own counterweight and no additional counterweight is needed. Rigid guides are provided for the cars which have shoes similar to those of ordinary lift cars.

In the space between cars there is a protective screen level with the front of the cars. This prevents people stepping into the shaft in between cars. It is still necessary to make sure that the landings and entrances are well illuminated. The

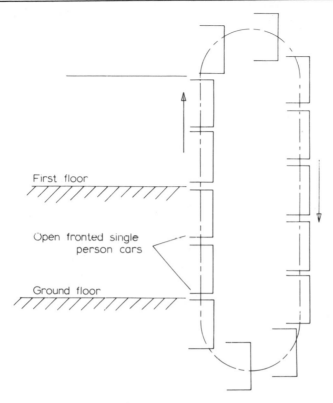

Fig. 172   Paternoster

cars are carried on a continuous steel link chain. The driving machinery is similar
to that of an escalator and is always placed above the well. It includes a brake
which is applied mechanically and held off electrically, so that the paternoster is
braked if the electrical supply fails. As in the case of both lifts and escalators
there is provision for handwinding.

A paternoster is started by a key operated switch, either at the ground floor
or at the main floor if this is other than the ground floor. There are emergency
stop buttons at each floor, in the pit and in the machinery space.

Although they have advantages, paternosters are not used very often. They
take up rather less space than escalators but have a lower carrying capacity. We
can show this by considering an escalator with a slope of $35°$ and a vertical rise
between treads of 230 mm. The distance along the slope between succeeding
steps is $230/\sin 35° = 400$ mm or 0.4 m. If the speed of the escalator is
0.5 ms^{-1}, the steps follow each other at intervals of $0.4/0.5 = 0.8$ s. As each step
can take one passenger, the carrying capacity is one person in every 0.8 s, or 75
persons per minute.

The vertical speed of a paternoster is at most 0.4 ms^{-1} and the car height cannot be less than 2.2 m. Neglecting any gap between the cars we see that the cars follow each other at intervals of $2.2/0.4 = 5.5$ s, and each car can take only one passenger. The carrying capacity is thus one person every 5.5 s, or 11 persons per minute.

However well constructed and carefully operated it is, a paternoster cannot help being more of a hazard than an escalator to the elderly, the infirm and above all to small children. This practically restricts its applications to industrial premises not needing a high carrying capacity.

Lifts, escalators and paternosters are covered by BS 2655.

17 Regulations

In most countries the supply of electricity is governed by legislation and we ought not to conclude this book without an account of the rules which apply in the United Kingdom. We have to refer to both the Electricity Acts and the Factories Acts, the latter of which apply only to industrial premises.

The Electricity Acts place an obligation on the Area Electricity Boards to provide a supply of electricity to everyone in their areas who asks for it. They naturally charge for the electricity and may make a charge for making the connection to their distribution system. This will depend on how much they have to extend that distribution network in order to reach the new consumer's premises. The Acts also confer power on certain government departments to make further regulations to control the supply of electricity.

Under these powers there have been made the Electricity Supply Regulations 1937. These deal chiefly with the standards of service and safety to be met by the Area Boards, and are not of direct concern to the designer of services in a building who is concerned with what happens on the consumer's side of the connection and not with what goes on in the road outside. The Regulations do however give the Area Board some powers of supervision over the consumer's installation. The chief of these is that the Board may not connect a consumer's installation to its supply if it is not satisfied that the insulation meets a prescribed value and that the installation has adequate protective devices. If a consumer does not comply with the regulations the Board may refuse to connect him, or if he has already been connected may disconnect him.

These regulations are somewhat general and it is conceivable that there could be doubt about their precise interpretation. In practice difficulties hardly ever arise, and there are probably two reasons for this. Firstly, Area Boards do not in practice inspect installations and are content with the installer's certificate of completion showing the insulation resistance. Secondly, and more importantly, installations complying with the Institution of Electrical Engineers' Regulations for the Electrical Equipment of Buildings are deemed to comply with the Electricity Supply Regulations. This brings us to consideration of the I.E.E. Regulations, and we can note a curious, and perhaps typically British, feature about them. They are the most important regulations which in practice have to be observed and yet they are made by a private body and have no legal force of their own. This comes about in the following way.

There is no legal need for an installation to comply with the I.E.E. Regulations. If an installation satisfies the Supply Regulations the law does not care whether or not it also satisfies the I.E.E. Regulations. But the law also says that if it happens to satisfy the I.E.E. Regulations it will be deemed to satisfy the Supply Regulations, and in practice this is the easiest way of showing that the Supply Regulations have been satisfied. As a result everyone in the industry is familiar with the I.E.E. Regulations, but very few people are aware of the curiously roundabout legal sanction behind them.

The 15th Edition of the I.E.E. Regulations was issued in March 1981. References to the Regulations in this book are to the 15th Edition.

Part 1 of the Regulations sets out fundamental requirements for safety. Part 2 contains definitions of terms. Part 3 lists the main features of an installation which have to be taken into account in applying the subsequent parts. Part 4 describes the measures to be taken for protection against the dangers that may arise from the use of electricity and Part 5 deals with the selection of equipment and accessories and with the details of construction and installation. Part 6 is concerned with inspection and testing and there are 15 appendices giving further details of methods of complying with the Regulations.

Chapter 13 of Part 1 of the I.E.E. Regulations states fundamental requirements for safety. If this chapter is not complied with it may be taken that the Area Board would be justified in disconnecting the supply. Parts 3 to 6 of the Regulations set out methods and practices which are considered to meet the requirements of Chapter 13. A departure from these parts of the Regulations does not necessarily involve a breach of Chapter 13 but should be given special consideration. In fact the Regulations make clear that they are not intended to discourage invention and that departure from them may be made if it is the subject of a written specification by a competent body or person and results in a degree of safety not less than that obtained by adherence to the Regulations. There are many passages in the 15th Edition of the I.E.E. Regulations which make it clear that it is perfectly in order for a competent Engineer to depart from the precise techniques described in the Regulations where he is dealing with an exceptional situation.

If a manufacturer wishes to introduce a new technique which is not envisaged in the current edition of the Regulations he can apply for a certificate from the I.E.E. Wiring Regulations Committee to the effect that the new technique is not less safe than those complying with the Regulations.

Much of what has been said in previous chapters is based on the methods and practices described in the I.E.E. Regulations and there seems little point in attempting either an abridgement or a gloss on the Regulations here. The Regulations do not contain instruction in the basic engineering principles on which they are based; they are regulations and not a textbook. An engineer who has followed a suitable course and understood the principles of electrical services should be able to read and understand the regulations without the interposition of a detailed commentary. A designer who follows the principles we have tried to explain in this book ought to find that his schemes almost inevitably comply with the regulations.

The other main source of legislation we have to refer to are the Factories Acts. These are concerned with safety in factories but do not themselves contain detailed rules for the use of electricity. Instead, they confer power on the Secretary of State to make Regulations. Such Regulations have been made and are known as the Electricity (Factories Acts) Special Regulations 1908 and 1944. It is the task of Factory Inspectors, who are government officials appointed by the Home Office, to see that these Regulations are observed. There are only 32 clauses in the Regulations and the wording is very general. Nevertheless it covers all likely sources of danger and is adequate to enable the Factory Inspectors to insist that all electrical services in factories are properly installed and maintained. Compliance with the I.E.E. Regulations will almost inevitably satisfy all the requirements of the Electricity (Factories Acts) Regulations, and the design principles explained in this book take into account the requirements of the Regulations.

18 Design Example

In order to illustrate the practical application of the principles discussed in previous chapters we shall, in this chapter, describe a typical industrial design. The example chosen is taken from a scheme handled in the author's office some years ago. The buildings of a disused factory were taken over by a chemical manufacturing company which proposed to adapt them as a new works. Electric services were needed for lighting and power to machinery.

The general plan of the buildings is shown in Figs. 173—175, which also show the main part of the lighting layout. As the design of lighting has been excluded from the subject matter of this book it is not proposed to reproduce the lighting calculations here, but it should be noted that after the number of lights needed in each area had been calculated they were positioned with regard to the layout of the machinery as well as to the need to maintain reasonable uniform levels of illumination.

The factory consists of an east building of two storeys with a basement and a three-storey west building with a covered yard between them extending the full height of the east building. There is a walled car park adjacent to the buildings and a new boiler-house was to be built in this area. Since the existing buildings provided more space than was needed for the new works part of the west building was to be left unoccupied: no services were to be installed in this part but the installation as a whole was to be capable of extension into this area.

The bulk of the lighting consisted of twin-tube 5 ft fluorescent fittings with some single-tube fittings in passages and areas requiring lower illumination. A few incandescent fittings were provided in toilets and on stairs (not all of which are shown in Figs. 173—175). The covered way between the occupied and unoccupied sections of the west building in which materials would be hoisted to the upper levels was lit by three wall-mounted mercury lamps at ground-floor level and three at second-floor level. One end of the west building contained tall machinery on the ground floor and the first-floor slab was not carried across this. An area of double the normal height was thus left and this was lit by wall-mounted mercury lamps at the lower level and high-bay industrial mercury

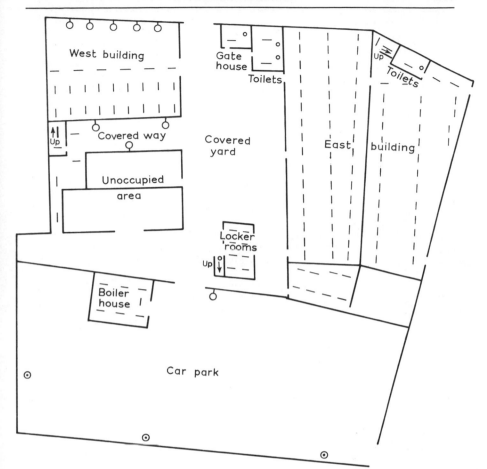

Fig. 173 Factory ground floor lighting layout

fittings under the first-floor ceiling. The covered yard was lit by wall-mounted high-pressure sodium floodlights at the level of the first-floor ceilings. Four street-lighting lanterns were provided for the car park, three of them being mounted on columns on the roadway from the building and one on a bracket on the wall of the building.

The first stage in the design was to arrange the lights in circuits and to arrange the circuits in convenient groups to be served from several distribution boards. The lighting would have to be divided in a suitable manner between the three phases to give as nearly as possible the same loading on all three phases and this had to be borne in mind when the lights were arranged into circuits. For convenience, the different types of fitting used were listed, as shown in Table 11.

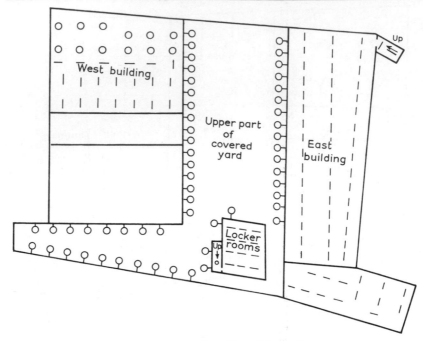

Fig. 174   Factory first floor lighting layout

It was decided that in this type of factory the lighting could be run in 2.5 mm² cable fused at 15 amps. To allow a margin for safety and small alterations the circuits would be designed to carry not more than 12 amps each. Although it was intended to use three different sizes of mercury lamp it was felt that there

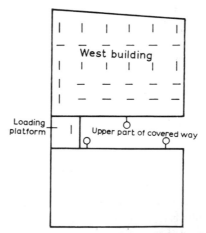

Fig. 175   Factory second floor lighting layout

Table 11

Ref.	Type	Current (amps)
A	5 ft twin fluorescent	0.92
B	5 ft single fluorescent	0.46
C	Wall-mounted 125 watt MBF	1.15
D	Tungsten bulkhead	0.42
E	High-bay industrial mercury with	
	250 watt MBF lamp	2.15
F	Wall-mounted area floodlight with	
	250 watt SON lamp	3.0
G	Bulkhead fitting with 50 watt MBF/U lamp	0.6
H	Side-entry street-lighting lantern with	
	35 watt SOX lamp	0.6

was a possibility that at some time in the future a works manager might change the fittings without checking the capacity of the wiring and it was therefore decided to design all the circuits serving fittings with mercury lamps to be capable of taking 250 watt lamps. Similarly, circuits serving single-tube fluorescent fittings would, where appropriate, be designed to take twin-tube fittings so that the fittings could at any time be replaced without alterations to the wiring. Hence, a maximum number of fittings on a circuit would be:

$$\frac{12}{0.92} = 13, \text{ but say 12 fluorescent fittings or}$$

$$\frac{12}{3.0} = 4 \text{ sodium fittings or}$$

$$\frac{12}{2.15} = 5.6 \text{ but, say, 4 mercury fittings}$$

It was clearly going to be desirable to control more than this number of lights from one switch and it was decided to do so by switching the lights through contactors. One switch would operate a multi-pole contactor controlling several lighting circuits.

At this stage a check was made on the voltage drop in the lighting circuits. Probable positions of distribution boards were guessed and from the drawings the average length of a lighting circuit was estimated as 35 metres. The maximum volt drop allowed on 240 volts is 6 volts and it seemed reasonable to allow half of this in the sub-mains and half in the final circuits, that is to say 3V. The volt drop of 2.5 mm^2 cable is 16 mV per ampere per metre.

$$3000\,(\text{mV}) = 16 \times I \times 35 \left(\frac{\text{mV}}{\text{A.m}} \times A \times m\right)$$

$$I = 5.35 \text{ amps}$$

Clearly the need to reduce voltage drop was more critical than the current rating of the cable and the number of fittings per circuit would have to be reduced. Acceptable figures would be 6 fluorescent fittings or 2 sodium or 2 mercury fittings per circuit.

Table 12

Area	Fitting				No. off circuit
	Ref.	No. off	Amps each	Amps total	needed
Gate-house	A	1	0.92	0.92	
	B	2	0.46	0.92	
	D	3	0.42	1.26	
				3.10	1
Pump house	A	8	0.92	7.36	2
Boiler house	A	7	0.92	6.44	2
Covered way	C	6	2.15	12.9	3 controlled by 1 contactor
W bldg Grd flr	A	24	0.92	21.00	3:1 switched directly, 2 controlled by 1 contactor
	G	5	2.15	4.30	2
E bldg Stores	A	48	0.92	44.2	8 controlled by 4 contactors
Ovens	A	33	0.92	30.36	6 controlled by 4 contactors
Toilet area	A	1	0.92	0.92	
	B	3	0.46	1.38	
	D	1	0.42	0.42	
				2.72	1
Maintenance	A	8	0.92	7.36	2
Lockers	A	8	0.92	7.36	2
Stairs	D	1	0.42	0.42	1
1st Floor					
Covered yard	F	34	3.0	102.0	16 controlled by 8 contactors
Side yard	F	20	3.0	60.0	10 controlled by 4 contactors
Lockers	B	12	0.46	5.52	1
W bldg 1st flr	A	20	0.92	22.08	6 controlled by 2 contactors
	E	12	2.15	25.80	6 controlled by 2 contactors
E bldg 1st flr	A	48	0.92	44.2	8 controlled by 4 contactors
1st flr					
Maintenance	A	7	0.92	6.44	2
1st flr					
Pump house	A	7	0.92	6.44	2
E bldg Cellar	A	12	0.92	11.04	2
W bldg 2nd flr	A	30	0.92	27.6	6 controlled by 2 contactors
W bldg 3rd flr	B	5	0.46	2.30	1

The loadings were now estimated for each area in a convenient tabulated form as shown in Table 12.

This formed a preliminary guide. The number of circuits in each area was decided by referring to the maximum number of fittings per circuit as determined above and also with an eye to convenient switching arrangements. At the same time, some margins were allowed to make it possible to adjust the circuit arrangements later without major modifications to the distribution scheme. It will be noted for example that the car-park lights are not included in the table. This was because the design had to proceed before the client had taken final decisions on all his requirements. The fact that last-minute alterations would certainly be made had therefore to be kept constantly in mind.

The total load from Table 12 was 460.94 amps. It should therefore be distributed to give about 150 amps per phase. An ideally equal distribution could not be hoped for but each phase should carry between 140 and 160 amps and at the same time each phase should be contained within a reasonably clear zone of the building. As a first step towards achieving this the loads for each area were summarized from Table 12, as shown in Table 13. They were then arranged in three

Table 13

Area	Load in amps
W bldg grd flr	25.30
Pump house	7.36
Gate-house	3.10
Covered way	12.90
	48.66
Boiler house	6.44
W bldg 1st flr	47.88
Covered yard (main area)	102.00
Covered yard (side area)	60.00
W bldg 2nd flr	27.6
W bldg 3rd flr	2.3
E bldg grd flr	44.20
E bldg grd flr	30.36
E bldg Toilet area	2.72
Maintenance	7.36
Lockers	7.36
Stairs	0.42
	92.42
E bldg 1st flr	44.20
E bldg Maintenance	6.44
E bldg Pump house	6.44
Lockers	5.52
	62.60
E bldg cellar	11.04

Table 14

Red phase	
W bldg 2nd flr	27.6 amps
W bldg 3rd flr	2.3
Covered yard main area	102.0
Boiler house	6.4
	138.3

Yellow phase	
W bldg grd flr	48.66
W bldg 1st flr	47.88
East cellar	11.04
Covered yard site area	60.00
	167.58

Blue phase	
E bldg grd flr	92.42
E bldg 1st flr	62.60
	155.02

groups for the three phases. After two attempts the results shown in Table 14 were obtained.

This was not as good as had been hoped for. However, the process of manipulating the figures had given the designer a feel for them and he realized that he was not likely to get any further improvement at this stage. It would be possible to make some adjustment after the distribution boards were scheduled and this was done next.

The lights and switching were shown on drawings. In each area, the fittings were grouped into circuits in accordance with the maximum number of fittings per circuit previously determined. Clearly the fittings on any one circuit must be in a reasonably compact group. Also, although the fittings on one circuit can be controlled by more than one switch, the converse is not true: one switch cannot control fittings on several circuits unless a multi-pole contactor is used. The most practicable way of settling these matters is to mark the circuits and switching groups on drawings of an adequately large scale.

Standard distribution boards are available with 12 and 16 ways. Suitable positions were chosen on the drawings for distribution boards to serve groups of 7 to 12 circuits to allow a reasonable number of spare ways on each board. The positions were chosen to keep the final sub-circuits reasonably short and so that as far as possible each board would be in the 'centre of gravity' of the area it was serving. It became evident in this process that the second and third floors of the west building should be served from a single board, that the gate-house would need its own board, that three distribution boards would conveniently handle both parts of the covered yard, that the ground and first floors of the east building would each need two distribution boards and that the cellar of the east building would be most conveniently served from the gate-house. The information from the drawings was then summarized in distribution-board schedules which are reproduced in Table 15.

Table 15   Lighting distribution boards

| Board No 1 | | E bldg maintenance area 1st flr | | |
| | | Phase Y   sub-main 35 mm^2 | | |

Circuit no.		No. and location of lights	Fuse (amps)	Cable (mm^2)
1	4	Pump house and changing rooms grd flr	15	2.5
2	4	Changing rooms grd flr	15	2.5
3	5	1st flr maintenance area	15	2.5
4	5	1st flr maintenance and sub-station	15	2.5
5	6	Pump house	15	2.5
6	6	Changing rooms 1st flr	15	2.5
7	5	Changing rooms 1st flr and stairs	15	2.5
8	4	Car park lights	15	2.5
9				
10				
11				
12				

| Board No 2 | | W bldg grd flr | | |
| | | Phase Y   sub-main 35 mm^2 | | |

Circuit no.		No. and location of lights	Fuse (amps)	Cable (mm^2)
1	5	Bulkheads on wall	15	2.5
2	3	Production area	15	2.5
3	3	Production area	15	2.5
4	4	Production area	15	2.5
5	3	Production area	15	2.5
6	4	Production area	15	2.5
7	4	Production area	15	2.5
8	3	Hoist yard, low level	15	2.5
9	3	Hoist yard, high level	15	2.5
10	5	Rear entrance	15	2.5
11	4	Stairs	15	2.5
12				
13				
14				
15				
16				

Table 15 *Continued*

Board No 3   W bldg 1st flr
                   Phase Y   sub-main 35 mm^2

Circuit no.	No. and location of lights		Fuse (amps)	Cable (mm^2)
1	2	Mercury over vats	15	2.5
2	2	Mercury over vats	15	2.5
3	2	Mercury over vats	15	2.5
4	2	Mercury over vats	15	2.5
5	2	Mercury over vats	15	2.5
6	2	Mercury over vats	15	2.5
7	3	Fluorescent production area	15	2.5
8	4	Fluorescent production area	15	2.5
9	4	Fluorescent production area	15	2.5
10	3	Fluorescent production area	15	2.5
11	4	Fluorescent production area	15	2.5
12	4	Fluorescent production area	15	2.5
13				
14				
15				
16				

Board No 4   W bldg 2nd flr
                   Phase R   sub-main 35 mm^2

Circuit no.	No. and location of lights		Fuse (amps)	Cable (mm^2)
1	4	Production area	15	2.5
2	5	Production area	15	2.5
3	5	Production area	15	2.5
4	4	Production area	15	2.5
5	5	Production area	15	2.5
6	5	Production area	15	2.5
7	3	Laboratory and landing	15	2.5
8	5	Third floor	15	2.5
9				
10				
11				
12				

Table 15 *Continued*

| Board No 5 | E bldg grd flr stores |
| | Phase B   sub-main 35 mm² |

Circuit no.	No. and location of lights		Fuse (amps)	Cable (mm²)
1	4	Stores	15	2.5
2	4	Stores	15	2.5
3	4	Stores	15	2.5
4	4	Stores	15	2.5
5	4	Stores	15	2.5
6	4	Stores	15	2.5
7	4	Stores	15	2.5
8	4	Stores	15	2.5
9	4	Stores	15	2.5
10	4	Stores	15	2.5
11	4	Stores	15	2.5
12	4	Stores	15	2.5
13				
14				
15				
16				

| Board No 6 | E bldg grd flr oven area |
| | Phase B   sub-main 35 mm² |

Circuit no.	No. and location of lights		Fuse (amps)	Cable (mm²)
1	5	Circulation area	15	2.5
2	4	Ovens	15	2.5
3	4	Ovens	15	2.5
4	4	Ovens	15	2.5
5	4	Ovens	15	2.5
6	6	Ovens	15	2.5
7	6	Ovens	15	2.5
8	5	Maintenance area	15	2.5
9	4	Maintenance area	15	2.5
10	5	Toilets and stairs	15	2.5
11				
12				
13				
14				
15				
16				

Table 15   *Continued*

| Board No 7 | | E bldg 1st flr stores | | |
| | | Phase B   sub-main 35 mm² | | |

Circuit no.	*No. and location of lights*		*Fuse (amps)*	*Cable (mm²)*
1	4	Stores	15	2.5
2	4	Stores	15	2.5
3	4	Stores	15	2.5
4	4	Stores	15	2.5
5	4	Stores	15	2.5
6	4	Stores	15	2.5
7	4	Stores	15	2.5
8	4	Stores	15	2.5
9	4	Stores	15	2.5
10	4	Stores	15	2.5
11	4	Stores	15	2.5
12	4	Stores	15	2.5
13				
14				
15				
16				

| Board No 8 | | Gate-house | | |
| | | Phase Y   sub-main 16 mm² | | |

Circuit no.	*No. and location of lights*		*Fuse (amps)*	*Cable (mm²)*
1	6	Lodge and toilets	15	2.5
2	6	E bldg cellar	15	2.5
3	6	E bldg cellar	15	2.5
4				

| Board No 9 | | Covered yard | | |
| | | Phase R   sub-main 35 mm² | | |

Circuit no.	*No. and location of lights*		*Fuse (amps)*	*Cable (mm²)*
1	2	Covered yard	15	2.5
2	2	Covered yard	15	2.5
3	1	Covered yard	15	2.5
4	2	Covered yard	15	2.5
5	2	Covered yard	15	2.5
6	1	Covered yard	15	2.5
7	2	Covered yard	15	2.5
8	2	Covered yard	15	2.5
9	2	Covered yard	15	2.5
10				
11				
12				

Table 15   *Continued*

Board No 10		Covered yard Phase R sub-main 35 mm²		

Circuit no.		No. and location of lights	Fuse (amps)	Cable (mm²)
1	2	Covered yard	15	2.5
2	2	Covered yard	15	2.5
3	2	Covered yard	15	2.5
4	2	Covered yard	15	2.5
5	2	Covered yard	15	2.5
6	2	Covered yard	15	2.5
7	2	Covered yard	15	2.5
8	2	Covered yard	15	2.5
9	3	Covered yard	15	2.5
10	3	Covered yard	15	2.5
11				
12				

Board No 11		Covered yard Phase Y sub-main 35 mm²		

Circuit no.		No. and location of lights	Fuse (amps)	Cable (mm²)
1	2	Covered yard	15	2.5
2	2	Covered yard	15	2.5
3	2	Covered yard	15	2.5
4	2	Covered yard	15	2.5
5	2	Covered yard	15	2.5
6	2	Covered yard	15	2.5
7	2	Covered yard	15	2.5
8	2	Covered yard		
9				
10				
11				
12				

A further table was then made in order to decide on which phase each of these boards should be and this is given in Table 16.

The figures in the three right-hand columns were entered in pencil, rubbed out and moved from column to column until by a process of trial and error quite a good balance over the phases was obtained. The first attempt was made on the basis of the provisional phasing decided on before the distribution boards had been scheduled.

The size of sub-main necessary to serve these boards was next calculated. The necessary current rating was evident from Table 16 but it was also necessary to calculate the size of cable needed to give an acceptable voltage drop. The distance from the intake to the furthest board was measured on the drawings and found

Table 16

Board no.	Area	Amps	Phase R	Y	B
1	E and W bldgs grd flr	24		24	
2	W grd flr	32		32	
3	W 1st flr	30		30	
4	W 2nd and 3rd flr	31	31		
5	E grd flr	48			48
6	E grd flr	45			45
7	E 1st flr	48			45
8	Gate-house	15		15	
9	Covered yard	48	48		
10	Covered yard	66	66		
11	Covered yard	48		48	
		Total	145	149	138

to be 98 m. This was rounded off to 100 m for the purpose of calculation. The current taken by the most distant board was 31 amps and for the calculation this was rounded off to 30 amps. It had previously been assumed that 3 volts would be lost in the final sub-circuits and it was now decided to allow a 2 volt drop in the sub-main. This would make the total less than the permissible maximum but there is no restriction on how low the voltage drop is and it seemed prudent to allow a margin for future extensions and also for possible alterations in the final positions of distribution boards and routes of cables.

2 volts = 2000 millivolts

Permissible drop is given by

$$2000 = \frac{mV}{A.m} \times 30 \times 100 = 3000 \times \frac{mV}{A.m}$$

$$\frac{mV}{A.m} = \frac{2000}{3000} = 0.67$$

35 mm^2 cable has a voltage drop of 1.2 mV/A.m and is rated at 90 amps.

A larger sub-main would seem unreasonable for the loads involved. Although the volt drop of 35 mm^2 cable is higher than the calculated figure, the calculation was on the safe side and was carried out only for the longest sub-main. The next size of cable is 50 mm^2 which is considerably harder to handle and therefore more expensive to install. It would be rather unusual to use such large cable for lighting distribution and it was therefore decided that 35 mm^2 cable would be acceptable. Each of these cables would be served from a 60 amp switch fuse. An exception was made for Board No 8 which would carry only 15 amps. By inspec-

Table 17

Item	No. off	H.P each	Running current (amps per phase)		
			Each	Total	Allow for diversity
'A' Agitator	4	5	8	32	16
Type 1 mill	1	50	70	70	35
Type 2 mill	1	25	36	36	—
'A' Mixers	2	5	8	16	8
Shakers	3	3	5	15	10
Extractor	13	5	8	104	52
'B' Mixers	2	15	22	44	22
'G' Mixers	10	20	30	300 ⎱	200
'G' Mills	10	10.5	16	160 ⎰	
Rotary valves (Single-phase motors)	20	0.25	3	60	15
'A' Pumps	1	7.5	11	11	—
'B' Pumps	1	5	8	8	8
'C' Pumps	3	4	6	18	9
'D' Pumps	1	5	8	8	8
'E' Pumps	3	7.5	11	33	11
'F' Pumps	3	7.5	11	33	11
Ovens	18	5	8	144	96
Hoist	1	5	8	8	—
'A' Fans	3	7.5	11	33	22
Conveyors	1	10	15	15	15
Dissolver	5	7.5	11	55	33
Coupling tanks	3	15	22	66	44
Lift	1	1.0	2	2	—
'B' Agitators	1	2	3	3	3
'G' Pump	1	3	5	5	—
'B' Fans	1	30	40	40	—
Boiler burner	1	10	15	15	15
'H' Pumps	1	15	22	22	22
'C' Fans	1	7.5	11	11	11
Burner auxiliary motor	1	3	5	5	5
'J' Pumps	1	5	5	5	5
		Total amps per phase		1369	676
		kVA over 3 phases		982	486

tion and without any calculation, it was decided that a 16 mm^2 cable rated at 53 amps with a volt drop of 2.6 mV/A.m would be adequate for this. It would be served from a 30 amp switch fuse.

Attention was now turned to the design of the power distribution. A list of the machinery to be installed was obtained from the client and written out as shown in Table 18. The locations of the equipment were also obtained and are shown in Figs. 176—178. It should be noted that all power equipment was to be 3-phase except for FHP motors on rotary valves. Most of it was accounted for by motors driving pumps, agitators and other mechanical equipment: the running currents per phase were taken from standard motor performance tables.

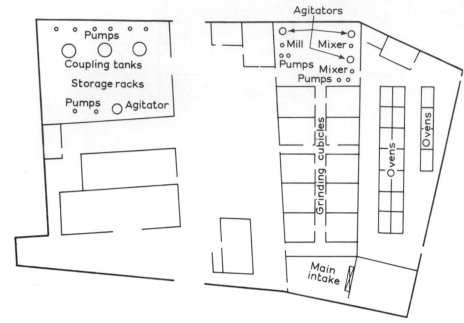

Fig. 176   Factory ground floor equipment layout

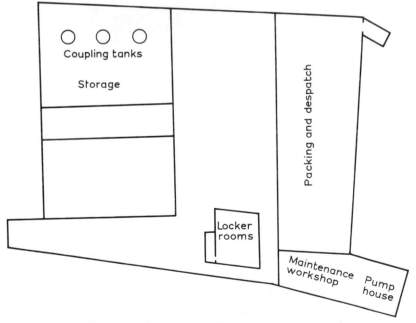

Fig. 177   Factory first floor equipment layout

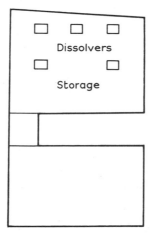

Fig. 178   Factory second floor equipment layout

The allowance for diversity was based on the designer's previous industrial experience and his assessment of what equipment might normally be in use simultaneously. The lighting load on the most heavily loaded phase was 149 amps and in view of the nature of the building it seemed reasonable to apply a diversity factor of 0.6 to this, giving an after-diversity lighting load of 90 amps. Addition of this to the after-diversity load of 766 amps per phase which is 570 kVA over all three phases. This could be conveniently catered for by 800 amp busbars at the main intake.

A difficulty arose over this figure. The supply to the existing board came from a 315 kVA transformer. If the electricity board were to be asked for a bigger supply they would make a substantial charge which the factory owner wished to avoid. The client also thought the calculated load was high but could not dispute the total installed load. He told the designer that at an older but similar works belonging to the same company measurements showed that the actual maximum demand was 27% of the total installed load. If the same figure were applied to the new factory the maximum demand would be 0.27 x (982 kVA power load + 120 kVA lighting load) = 298 kVA which would be within the capacity of the existing supply. The client therefore wanted this figure to be used. Whilst unable to challenge the client's measurements the designer felt that a diversity factor of 27% was surprisingly low. He pointed out that if the distribution was designed on this figure and it turned out to be low it would be very difficult and expensive subsequently to increase the capacity of the installation. He was reluctant to work on this basis. After discussion it was agreed that 800 amp busbars would be installed at the main intake but would be served through a 400 amp switch fuse from the existing 315 kVA (equivalent to 440 amps per phase) supply. This would make it possible to cater for a larger load if the need arose without expensive alterations but would not increase the initial cost very much. It therefore satisfied both points of view.

A description such as this inevitably makes the design process seem very precise whereas in practice at each stage there are many unknown facts for which the designer has to make a guessed allowance. In the present case the plant design was proceeding at the same time as the electrical design and neither the ratings nor the positions of all the equipment were finally settled. Table 17 is in fact based on the third attempt to draw up such a list; it would be an unnecessary waste of space to reproduce the earlier tables which differed only in detail. However the element of uncertainty led to two important decisions about the general scheme.

Firstly, it was decided to use busbars with separately mounted switch fuses rather than a cubicle-type switchboard. This would give excellent flexibility for future extensions and also for changes and additions which might become necessary before the installation was completed. It seemed quite likely that this would be necessary because of the uncertainty of the final plant layout.

Secondly, it was decided that the design of the power installation would go only as far as the final distribution boards. The final sub-circuits from these to the various motors would be settled on site after the machines were installed. In areas where there was to be a lot of equipment horizontal busbars could be run along the building walls with tap-off boxes spaced as required.

With these considerations in mind the load was listed again but this time area by area. The load was added for each area and a decision made on the size and rating of the distribution board to serve that area. This list is shown in Table 18. It will be noticed that some additional items not listed in the previous table were added at this stage.

Table 18

Area	Item	No. off	Running current (amps per phase) Each	Total	After diversity
1	'A' Agitators	2	8	16	
E bldg	Type 1 mills	1	70	70	
N end	'B' Pumps	1	8	8	
	'C' Pumps	1	6	6	
	Shaker	1	5	5	
		6		115	100
2	'A' Agitator	2	8	16	16
E bldg	'B' Mixer	2	22	44	44
N end	Type 2 mill	1	36	36 ⎫	
	'A' Mixer	2	2	16	
	'C' Pump	2	6	12 ⎬	41
	Shaker	2	5	10	
	Extractor	1	8	8 ⎭	
		12		142	101

Table 18 (continued)

Area	Item	No. off	Running current (amps per phase)		
			Each	Total	After diversity
3	per cubicle				
E bldg	'G' Mixer	1	30		
Grinding	'G' Mill	1	16		
Cubicles	Rotary valve	2	6		
	Extractor	1	8		
		5	60		
4	Ovens	18	8	144	
E bldg	Extractors	2	8	16	
Ovens				160	
5	Hoist	1	8	8	
W bldg	'A' Fans	3	11	33	
3rd flr	Conveyors	1	15	15	
	'D' Pump	1	8	8	
		6		64	45
6	Dissolver	5	11	55	33
W bldg					
2nd flr					
7	Coupling tanks	3	22	66	22
W bldg					
1st flr					
8	'E' Pump	3	11	33	
W bldg	'F' Pump	3	11	33	
Grd flr	Lift	1	2	2	
	'B' Agitator	1	3	3	
	'A' Pump	1	11	11	
	'G' Pump	1	5	5	
		10		87	40
9	'B' Fan	1	40		
Boiler	Burner	1	15		
Room	'M' Pump	1	22		
	'C' Fan	1	11		
	Auxiliary motor	1	5		
	'J' Pump	1	5		
		6	98		

The load in areas 1 and 2 could be catered for by a 24 way 300 amp TPN distribution board served from a 200 amp switch fuse.

A check with manufacturers' catalogues showed that it would not be possible to get a standard board with outgoing fuse-ways in the wide range of sizes needed, that is to say from 5 amps to the 100 amps needed for the mills. It would therefore be necessary to use two separate boards, one with fuse-ways from 2 to 30 amps and one with fuse-ways from 20 to 100 amps. The former would have fourteen items with an installed load of 97 amps giving about 58 amps after diversity and the latter would serve four items with an installed load of 150 amps giving about 90 amps after diversity. A 20 way 60 amp TPN board and a 6 way 100 amp TPN board would meet these requirements.

In area 3, the equipment in one cubicle only is listed. The mixer and the mill do not run at the same time. The rotary valves run intermittently, therefore the maximum simultaneous demand can be assessed as 30 + 3 + 8 = 41 amps. There are ten such cubicles making the installed load 10 x 60 = 600 amps. After diversity this will be say 400 amps.

In this area there is a total of 10 x 5 = 50 pieces of equipment. To allow spare capacity 65 ways on a distribution board are needed.

Each bay is approximately six metres long. A busbar would have not more than 6 tap-off points along this length but each bay contains two cubicles with 10 pieces of equipment. Therefore a busbar is not the most practicable method and distribution boards should be used.

The distribution boards will be mounted on a wall. As there is a central gang-way midway between the two facing walls the arrangement will have to be symmetrical so that either 2 or 4 distribution boards will have to be used. This gives the possibility of either 2 off 36 way 300 amp TPN boards from 200 amp switch fuses or 4 off 18 way 200 amp TPN boards from 150 amp switch fuses.

At first sight the first alternative appeared cheaper but on checking with the manufacturers it was revealed that standard boards are not made as large as this so the second alternative had to be adopted.

Area 4 can be conveniently served by 200 amp TPN busbars fed by a 200 amp isolator at the busbars which can in turn be served from a 150 amp switch fuse on the main panel.

At this stage it had been decided that the east building would require 6 distribution boards and one set of busbars. All this could conveniently come from a subsidiary distribution centre in the building. The sum of the after-diversity loads calculated for this was 741 amps but allowing for diversity between the boards the maximum load on the busbars would be less. To allow for adequate short-circuit strength and also for future extensions it was decided to use 800 amp busbars for the subsidiary centre. It had already been decided, as explained above, that the main intake would have 800 amp busbars with a 400 amp incoming switch fuse. To give discrimination, the outgoing switch fuse could not be larger than 300 amps. The after-diversity load was probably still being over-estimated but whereas switch fuses and if necessary cables can be changed later it would be very expensive to replace the busbars. Therefore the local busbars can still be 800 amps but should have an incoming isolator of a lower rating. There is a fuse at the outgoing end of the cable from the main intake and there is no need for another fuse at its other end. As the fuse is 300 amps the isolator which is protected by the fuse should have a higher rating, say 400 amps.

Reference was made in the last paragraph to short-circuit strength. In fact no separate calculation was made for this design but the results of calculations on other projects were made use of.

The busbars are rectangular copper bars supported at regular intervals. The dimensions of the bars and the spacing of supports are given in the manufacturer's catalogue. If the length between two successive supports is treated as a simply-supported beam the maximum permissible bending moment can be calculated from the bending stress formula:

$$\frac{M}{I} = \frac{p}{y}$$

where M = bending moment (Nm)

I = second moment of area (m^4)

p = stress (N m^{-2})

y = distance from neutral axis to outermost fibre (m)

I and y are easily claculated for a rectangular section, p is the maximum allowable working stress of pure copper, and M is the moment to be calculated

The bending moment for a simply-supported beam with uniform loading is

$$M = \frac{wL^2}{8}$$

where M = bending moment (Nm)

w = load per unit length (N m^{-1})

L = distance between supports (m)

As M has been established and L is known, this enables w to be calculated to give the maximum permissible uniform load on each bar.

When a current flows in two parallel rectangular bars the resulting mechanical force between them is given by

$$w = \frac{120\, i^2 \times 10^{-8}}{s}$$

where w = force per unit length (N m^{-1})

i = current (amps)

s = spacing between bars (m)

If the force per unit length is taken as the maximum permissible uniform load which has just been calculated, and the spacing between the bars is known from the manufacturer's catalogue, this formula allows the maximum allowable value of the current to be calculated. This value is then the maximum current which the bars will be just strong enough to withstand and they should not be exposed to a possible short-circuit current higher than this.

Perhaps this seems a lengthy and somewhat circuitous piece of reasoning. It is, however, a typical example of the way in which the various requirements for a distribution system have to be fitted together. It has not been written as a description of the final scheme but rather to show the process by which the scheme was arrived at.

The subsequent distribution centre in the east building could also serve a distribution board in the maintenance area and the unit heaters for the space heating. No information was available at this stage of the equipment which would be installed in the maintenance workshop but a 12 way 60 amp TPN board would certainly be adequate. The area would have ten unit heaters which could be served from a 12 way SPN board.

At this stage there was still some uncertainty about the exact positions of the equipment in the lower part of the west building and indeed about how much of the equipment planned would be installed initially and how much left for the future.

Partly for this reason and partly to give the greatest possible flexibility for the final connections, it was decided after discussion with the client to provide a busbar under the first floor ceiling of the west building to serve the first and ground floors. It will be remembered that part of the first-floor slab was omitted to give a two-storey height to part of the ground floor. This made it possible to serve the ground floor from a busbar at high level on the first floor. Indeed in view of the lack of information available to the electrical designer about the height of the motors on the machinery to be installed this seemed the only reasonable thing to do.

From Table 18 the total amps per phase on the ground and first floors were 153 installed and 62 after diversity. 200 amp busbars served by a 200 amp switch fuse would be ample for this load (this is the lowest standard rating for busbars).

The second floor needed an 8 way 60 amp TPN distribution board served from a 60 amp switch fuse.

The third floor needed an 8 way 60 amp TPN distribution board served from a 60 amp switch fuse.

The loading for the various areas of the west building was then summarized as shown in Table 19. It became evident that the whole of this could conveniently come from a subsidiary distribution centre within the west building. Adequate margins and provision for future additions suggested that a suitable size would be 400 amp busbars served from a 200 amp switch fuse.

The space heating of the west building was to be provided by four unit heaters on the second floor and four on the first floor which all required a supply for fans and thermostats. Much of this equipment could be conveniently served from a 6 way SPN distribution board on its own floor and these two boards could also be served from the subsidiary distribution centre.

The only remaining area to be dealt with was the boiler room. This is Item 9

Table 19   Summary for W building

Floor	Running current (amps per phase)	
	Total installed	After diversity
Grd	87	40
1st	66	22
2nd	55	33
3rd	64	45
	272	140

in Table 18. There would be little diversity here and it was therefore decided to provide a 150 amp TPN distribution board served from a 150 amp switch fuse.

To save a multiplicity of sub-mains cables, it was decided that in the west building the lighting distribution boards would be served from the subsidiary distribution centre. The east building was nearer the main intake so that it would not be so cumbersome and expensive to run several cables between them. Also the number of switch fuses required on the east building distribution centre for the power boards alone was already quite high. It was therefore decided that the east building lighting boards would be served directly from the main intake.

The distribution scheme was now sketched, as shown in Fig. 179. This is the most convenient method of summarizing the decisions taken so far and checking for any inconsistencies or omissions. In its final form it is also the clearest way of explaining the scheme to be installed to contractors and suppliers.

It now remained to decide the sizes of the various sub-main distribution cables which, at this stage had not been written into the scheme of Fig. 179. The necessary current ratings were clear from the switch fuse ratings needed on the scheme but the cables also had to be calculated for voltage drop. It is only necessary to make sure that the voltage drop is not excessive and one of a limited number of standard-size cables must be chosen. The calculation was therefore simplified by taking 100 metres as the longest run of a sub-main cable and using the same length in calculating all of them. It had earlier been assumed that, of the total permissible voltage drop half would be in the final sub-circuit and half in the sub-mains. To keep to this assumption the sub-mains had to be calculated for a drop of 3 volts.

In view of the number of sub-mains involved the calculation was again set out in a tabular form as shown in Table 20. Column 2 gives the current rating. In column 3 the drop in millivolts per ampere metre to give a drop of 1 volt over the assumed total length of 100 metres has been calculated. This figure was then multiplied by 3 and rounded off in column 4 to give the millivolts per ampere metre for a 3 volt drop. Columns 2 and 4 thus showed the minimum requirements of the cable: the cable chosen to match this was then entered in columns 5, 6

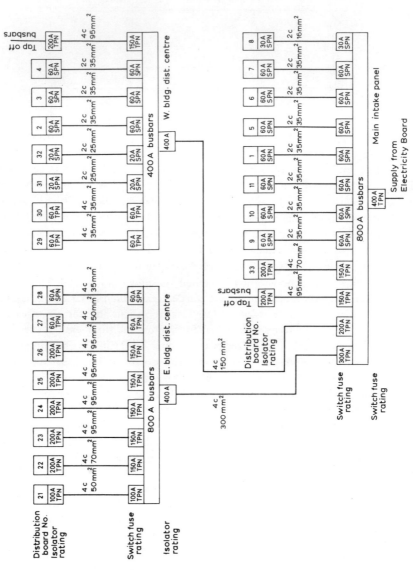

Fig. 179 Factory distribution diagram

Table 20

1	2	3	4	5	6	7
Board No.	Amps	mV/A.m for 1 volt drop $= \dfrac{1000}{amp \times 100}$	mV/A.m for 3 volt drop	Cable mm²	amp	mV/A.m
21	60	0.166	0.5	50	125	0.81
22	100	0.1	0.3	70	155	0.57
23	150	0.067	0.2	95	190	0.42
24	150	0.067	0.2	95	190	0.42
25	150	0.067	0.2	95	190	0.42
26	150	0.067	0.2	95	190	0.42
27	60	0.166	0.5	50	125	0.81
28	30	0.333	1.0	35	72	1.0
29	45	0.222	0.67	35	72	1.0
30	25	0.4	1.2	35	72	1.0
31	12	0.8	1.8	25	62	1.5
32	12	0.8	1.8	25	62	1.5
33	100	0.1	0.3	70	155	0.57
W. Busbars	100	0.1	0.3	70	155	0.57
Oven Busbars	150	0.067	0.2	95	190	0.42
E. Sub-centre	300	0.03	0.09	300	390	0.18
W. Sub-centre	120	0.08	0.24	150	250	0.29

and 7. The cable sizes were then entered on the scheme which is reproduced in Fig. 179 and this completed the design.

In this example the designer's freedom was restricted by the limitations of an existing building. In principle, where a new building is being designed, the designer of the electrical services can be called in early enough to suggest arrangements which would result in a more economical services installation. The effect of voltage drop or cable sizes makes it desirable to have the spaces allocated for intake panels and distribution boards as near as possible to the centre of the area being served. The provision of false ceilings, and horizontal and vertical ducts influences the type of wiring system to be employed. The thickness of plaster may determine whether or not cables can be buried within it. If walls are built of a single thickness of brick it will not be practicable to chase them for cables or conduits, and the electrical installation may have to be run on the surface.

All these matters can be discussed by the electrical services designer and the architect at a very early stage and should in theory influence the building design. In practice it seems that since it is always possible to adapt the electrical installation to any building, purely architectural considerations always override the engineering ones. Many architects have no objection to conduit on the surface of walls even in a completely new building, and if this is accepted the type of construction no longer matters.

In the early stages of design the architect's ideas are very fluid and it is difficult for the electrical designer to make suggestions which are more than vague generalities. By the time he receives drawings on which he can start design work of his own the shape and style of the building have been settled and can no longer be altered to accommodate or simplify the services.

Thus the engineering designer's influence on the overall design of the building tends to exist more in theory than in practice.

When the work is to be put out to competitive tender it is necessary to draw up a specification describing the quality and standard of the equipment to be used and the standard of workmanship expected. The specification which was used for the scheme described in this chapter is reproduced in the following pages.

It is often prudent to include in a specification descriptions of equipment which may not be needed for the scheme as designed. Variations are frequently made during construction, and they could introduce a piece of equipment not originally needed. If it has not been described in the specification a short variation instruction can give rise to different interpretations which could result in a contractual dispute.

For example, the original scheme may not require any isolators as opposed to fuse switches and switch fuses. If an isolator is subsequently required and there is no clause in the specification covering isolators, the variation instruction must include a complete description or there is the possibility that an unsuitable type will be supplied. Since there is invariably less time available for drafting variations than for the original specification, it is better to have a few extra clauses in the specification than to risk contractual difficulties later on.

Specification of electrical works

1. *Cubicle panels*

Main and distribution switchboards as listed in the schedules shall be purpose-made and shall consist of a sheet-steel cubicle designed for access from front only or from front and rear according to site position, containing distribution busbars, all cable terminations and interconnections. All items of switch and fusegear are to be flush-mounted on the front of the cubicle which is to be finished grey stoved enamel.

Switchboards shall be suitable for controlling medium-voltage supplies and shall be comprehensively tested before despatch from the manufacturer's works to ensure satisfactory operation of all component parts. The tests shall include continuity and 2 kV flash tests. Busbar systems in switchboard shall be tested at 50 kA for 1 second.

Full protection shall be provided by means of mechanical interlocks with the covers and operating levers. The switchboard shall have an earthing terminal and earthing bar.

External connections shall be provided to allow all outgoing cables to terminate at either top or bottom of the switchboard.

2. *Busbars*

Busbar panels as shown on the drawings and listed in the schedules are to be of the ratings indicated and are to be of high-conductivity copper mounted on robust vitreous porcelain insulators complete with all necessary clamps. The busbars are to be enclosed in a stove-enamelled sheet-steel casing with cast-iron frame members and detachable top, bottom and side plates. The covers are to be of the screw-on type. The panels are to be provided with all necessary holes and bushes for incoming and outgoing cables.

Incoming and outgoing switch fuses of the types and ratings shown on the drawings and schedules shall be provided and installed adjacent to the busbar chamber. All necessary interconnections between switchfuses and busbars shall be made and all the equipment shall be fixed on a common angle-iron frame which is to be supplied as part of the electrical contract. The frame is to be painted one coat primer and two coats grey finish.

Where necessary trunking shall be supplied and installed from busbars to meter positions.

3. *Fuse switches*

Fuse switches shall be heavy-duty pattern fitted with HRC fuse links. They shall have heavy-gauge steel enclosures with cast-iron frame members, rust-protected and finished grey stoved enamel. Front access doors shall be fitted with dust-excluding gaskets and shall be interlocked so that they cannot be opened when the switch is 'on'. Operating handles shall be lockable in both the on and off positions. The top and bottom endplates shall be removable.

Each fuse switch shall be supplied complete with the correct HRC fuse links.

Each fuse switch shall have flag on-off indication.

Fuse switches shall be 500 V rating and shall be clearly marked with their current rating.

4. *Switch fuses*

Switch fuses shall be industrial pattern dust-proof type with HRC fuse links. They shall have enclosures fabricated from sheet steel finished grey stoved enamel with removable top and bottom endplates and shall have doors fitted with dust-proof gaskets. They shall have front-operated handles with visible on-off indication.

The interiors shall have vitreous porcelain bases fitted with plated non-ferrous conducting components. Switches shall be of the quick make-and-break type and have removable shields over the fixed contacts and removable moving contact bars.

Each switch fuse shall be supplied complete with the correct HRC fuse links.

Switch fuses shall be 500 V rating and shall be clearly marked with their current rating.

5. *Isolators*

Isolators shall be heavy-duty pattern with steel enclosures having cast-iron frame members, rust-protected and finished grey stoved enamel. Front access doors shall be fitted with dust-excluding gaskets and shall be interlocked so that they cannot be opened when the switch is on. Operating handles shall be lockable in both the on and off positions and shall have visible on-off indication.

Isolators shall be 500 V rating and shall be clearly marked with their current rating.

The moving contact assemblies are to be removable for inspection and maintenance.

6. *Distribution boards*

All distribution boards shall be single-pole or triple-pole with neutral bar, and with fuseways 20 or 30 ampere or above as specified.

All distribution boards shall be of the surface pattern in heavy sheet-steel cases of the 500 volt range with HRC fuse carriers.

On all triple-pole and neutral distribution boards the number of neutral terminals to be provided shall be the same as the total number of fuseways in the board. This information must be given to the manufacturers when ordering, together with information with regard to composite boards having a multiplicity of fuse ratings as specified.

In the case of flush installations, the distribution boards shall be mounted over flush adaptable iron boxes into which the conduits and wiring of the system will terminate. The boxes shall be of a size to be agreed on site with the consulting engineer.

Doors of all distribution boards shall be lockable either by means of a barrel-type lock with detachable key or by means of a modified door-fixing screw, bracket and padlock.

Ample clearance shall be provided between 'live' parts and the sheet-steel protection to allow cables to be brought to their respective terminals in a neat and workmanlike manner.

To separate opposite poles a fillet of hard incombustible insulating material shall be provided of sufficient depth to reach the inside of the door.

In each distribution board spare fuse carriers shall be provided and held in place by a suitable clip so that the carrier cannot be inadvertently dislodged.

Distribution boards shall be fixed at a height of two metres (6 ft 6 in) above floor level as measured from the bottom of the casing, or as directed by the consulting engineer.

7. *Starters*

A starter shall be provided for each motor as indicated on the drawings and schedules.

The starters shall be surface-mounted with a sheet-steel case containing a triple-pole contactor with vertical double break per pole having silver-faced contacts. They are to have continuously rated operating coils with inherent undervoltage release. Operating coils are to be supplied from phase to neutral. The starters shall have magnetic-type overload relays with adjustable oil dashpot time lags and stop/reset push-buttons in the front cover. The cover is also to contain the start push-button.

Starters are to have single-pole auxiliary switches and shall incorporate single-phase protection.

Star-delta starters shall have a time-delay device complete with all main and control wiring and terminal block for incoming and outgoing cables. The time-delay device shall be of the pneumatic pattern with an instantaneous reset allowing restarting immediately after a star-delta switching operation. The star and delta contactors are to be mechanically and electrically interlocked.

The overload relays are to be correctly set to ensure adequate protection without nuisance tripping.

8. *Steel conduit*

All steel Class 'B' conduits and conduit fittings throughout the whole of this installation shall comply in all respects with British Standard Specification BS 31:1940.

All PVC insulated cables, other than flexibles, shall be protected throughout their length with heavy-gauge screwed welded conduit (enamelled or galvanized as required) with the necessary malleable iron loop-in, draw-in, angle and outlet boxes. No type of 'elbow' or 'tee' will be allowed on works under this Specification.

Where adaptable boxes are used they shall be of cast iron or heavy-gauge sheet steel of not less than 12 gauge.

No conduit of less than 20 mm diameter shall be used.

A solid coupling shall be inserted in every flush conduit run at the point where it leaves a ceiling, wall or floor for ease of dismantling if required.

Except where otherwise stated conduit is to be finished black enamel.

No conduit shall be installed with more than two right-angle bends without draw-in boxes and draw-in boxes shall not be more than 8 metres apart.

All conduits, except where otherwise specified, shall drop not rise to the respective points. In no circumstances shall the conduit be erected in such a manner as to form a U without outlet, or in any other way that would provide a trap for condensed moisture.

Provision shall be made for draining all conduits or fixtures by a method approved by the consulting engineer.

No ceiling looping-in point box shall be used as a draw-in box for any other circuit than that for which such point box is intended.

Ceiling point boxes are to be of medium pattern malleable iron, with fixing holes at 2 in centres and conforming to BS Specification.

Flush ceiling point boxes which do not finish flush with the finished surface of the ceiling, etc. shall be fitted with malleable iron extension rings.

Horizontal or diagonal runs of flush conduit on structural or partition walls will not be permitted. All flush conduits shall drop or rise vertically to their respective points.

Connections between conduits and trunking and conduit and steel boxes, or between conduit and steel cases of distribution gear or equipment, shall be made by means of a flanged coupling and brass smooth-bore entry bush. The lead washer shall be fitted on the inside of the trunking or box, etc.

All lids for draw-in boxes, etc., whether of the BS or adaptable type shall be of heavy cast-iron or 12 gauge sheet steel, and shall be fixed (overlapping for flush work) by means of two or four 2 BA round-headed brass screws as required.

Conduits set through walls will not be permitted. When change of direction is required after passing through a wall an appropriate back outlet box is to be fitted.

All joints between lengths of conduit, or between conduit and fittings, etc. are to be threaded home and butted.

Sets and bends are to be made without indentation, and the bore must be full and free throughout. All screw-cutting oil must be carefully wiped off before joining up.

Conduit runs, as far as possible, are to be symmetrical and equally spaced.

The electrical contractor must take all precautions in situations likely to be damp to see that all conduits and boxes in the vicinity are rendered watertight.

During the progress of the work all exposed ends of conduits shall be fitted with suitable plastic or metal plugs. Plugs of wood, paper and the like will not be acceptable as sufficient protection.

Lighting, heating, power and any other types of circuit shall be run in separate conduits and no circuit of any one system shall be installed in any conduit or box of any other system.

The proposed runs shall be submitted to the consulting engineer for approval before work is commenced.

9. *Conduit fittings*

All conduit fittings shall be of malleable iron which shall conform to the British Standards Specification BS 31:1940.

All fittings shall be of the screwed pattern, and no solid or inspection elbows, tees or bends shall be installed. Generally, all conduit fittings shall be stove-enamelled black or other approved finish inside and out, but where galvanized conduit is installed, all fittings shall be galvanized by the hot process both inside and out. Such fittings shall be of Class B pattern.

All conduit fittings not carrying lighting or other fittings shall be supplied with suitable cast-iron covers with round-head brass screws. Where flush boxes are installed the covers shall be of the overlapping rustproof pattern.

All ceiling point boxes, except in the case of surface conduits, shall finish flush with the underside of the ceiling, extension rings being used where necessary.

Every flush ceiling point box to which a lighting fitting is to be attached shall be fitted with a break-joint ring of approved type.

Where surface conduit is used in conjunction with distance saddles, special boxes shall be used, to obviate the setting of conduit when it enters or leaves the boxes.

All conduit boxes, including boxes on and in which fittings, switches and socket outlets are mounted, shall be securely fixed to the walls and ceilings by means of not less than two countersunk screws, correctly spaced, and the fixing holes shall be countersunk, so that the screw heads do not project into the box.

10. *Flexible conduit*

Connections to individual motors and heating equipment run in conduit shall be made using a minimum of 300 mm of watertight flexible conduit. The conduit shall be Kopex LS/2.

Flexible conduit connecting to heating equipment shall employ butyl rubber insulated CSP sheathed cables and suitable terminal blocks shall be used in all boxes where a change in cable type is involved.

Earth continuity of all flexible conduits shall be maintained by $4\,\text{mm}^2$ minimum copper conductors forming one of the cores of the cable.

Flexible conduits shall be terminated with the Kopex couplings and connectors specially made for the purpose.

11. *Cable trunking*

Cable trunking shall be supplied and installed complete with fittings and accessories and shall be of an approved manufacture. It shall be manufactured from zinc- or lead-coated sheet steel finished stove enamelled grey or galvanised and shall be of 18 swg for sizes up to and including 75 mm x 75 mm section and 16 swg for sizes above.

All bends, tees, reducers, couplings, etc., shall be of standard pattern: where it is necessary for a special fitment to be used, it is to be fabricated by the manufacturers.

Where it is necessary to provide additional trunking over fixings, these shall be as supplied by the manufacturer and shall be applied with the manufacturer's special tools.

Where holes or apertures are formed in the trunking for cable entry, they shall be bushed with brass smooth-bore entry bushes, or PVC grommet strips. Cable supports are to be inserted in vertical runs of trunking and cables are to be laced thereto in their respective groups.

Fire barriers of hard insulating material shall be provided in vertical runs of trunking where they pass through floors.

Where more than one service is involved multi-compartment trunking shall be employed to separate the services.

12. *Cable tray*

Cable tray is to be made of 16 gauge perforated mild-steel sheet and is to be complete with all coupling pieces and bends, offsets and fixing brackets, to enable the tray to fit the structure accurately.

Where cables are taken over the edge of the tray they shall be protected by rubber grommets.

13. *MICC cable*

Mineral-insulated metal-sheathed cables shall be high-conducting copper conductors embedded in magnesium oxide and sheathed with copper with an overall covering of PVC.

All cable terminations shall be protected and sealed with ring-type glands with screw-on pot-type seals utilizing cold plastic compound and neoprene sleeving all of an approved pattern, and applied with the special tools recommended by the cable manufacturers.

Cold screw-on pot-type seals shall be used except where the ambient temperature in which the cable will operate will exceed 170°F, where the hot-type seal shall be used.

Four-core MICC cables shall *not* be used for ring circuits.

Vibration-absorbing loops shall be formed in MICC cables connected to motors, and other vibrating equipment.

Connections and joints in MICC cables shall only be made at the terminals of switches, ceiling roses, or at connector blocks housed in outlet boxes. Connector blocks shall have a minimum of two screws per conductor.

Where entry is made into equipment which does not have a spouted entry, the cable shall be made off by means of coupling, male bush, and compression washer.

MICC cables shall be neatly installed and shall be clipped by means of copper saddles secured by two brass screws. Where two, three or four cables are run together multiple saddles shall be used.

MICC cables shall be delivered to the site with the manufacturers' seals and identification labels intact and shall be installed in accordance with the manufacturers' recommendations and using the specialized tools recommended by the manufacturer. They shall be tested when installed before being sealed and again at the end of the contract.

They shall be sealed against the ingress of moisture at all times during the contract.

14. *PVC cable*

Cable PVC-insulated only, and PVC-insulated PVC-sheathed cable shall be 600/1000 volt grade to BS 6004:1969.

The cable shall be delivered to site on reels, with seals and labels intact and shall be of one manufacturer throughout the installation.

The cable shall be installed direct from the reels and any cable which has become kinked, twisted or damaged in any way shall be rejected. The installation shall be wired on the loop-in system i.e. wiring shall terminate at definite points (switch positions, lighting points, etc) and no intermediate connections or joints will be permitted. Cables shall not pass through or terminate in lighting fittings.

Where it is necessary to make direct connection between the hard wiring and flexible cord, this shall be done by means of porcelain-shielded connectors with twin screws. No lighting fitting shall be connected directly to the hard wiring.

The terminations shall be suitable for the type of terminal provided and shall be either sweated lugs of appropriate size, or eyelet or crimped type cable terminations, all of reputable manufacture. Shakeproof washers shall be used where electric motors are connected.

Where cable cores are larger than terminal holes, the cables shall be fitted with thimbles. For all single connections, they shall be doubled or twisted back on themselves and pinching screws shall not be permitted to cut the conductors. Cables shall be firmly twisted together before the connection is made.

In no circumstances shall cables be trapped under plain washers as a termination.

Cables shall be coloured in accordance with I.E.E. Regulations

Only two cables shall generally be bunched together at one terminal. In exceptional cases three cables may be bunched together at one terminal with the authority of the engineer given on site.

15. *Flexible cords*

All flexible cords shall comply with British Standard Specification BS 6500:1969, and shall consist of high-conductivity tinned copper conductors of the required cross-sectional area insulated and sheathed as detailed hereunder:—

Lighting pendants
2 core 0.75 mm² heat-resisting circular flexible cord EP rubber-insulated CSP-sheathed. Colour of sheath, white.

Heating apparatus and equipment requiring flexible cable connection
Heat-resisting circular flexible cable. EP rubber-insulated CSP-sheathed having the number of cores with cross-sectional areas as specified.

Apparatus and equipment, other than heating, requiring flexible cable connection
Circular flexible cable PVC-insulated PVC-sheathed having the number of cores with cross-sectional areas specified.

The cores of all flexible cords shall be coloured throughout their length and colour-coded to comply with the British Standard Specification.

16. *PVC SWA cable*
PVC-insulated single-wire armoured cables shall be 500/1000 volt grade and shall comply with BS 6346:1969.

The cable shall comprise round or shaped conductors, of equal cross-sectional area, composed of high-conductivity plain annealed copper wire insulated with PVC, coloured for identification. The cores to be laid up circular and sheathed with PVC. The cable shall be served with one layer of steel-wire armour and sheathed overall with PVC.

The cable shall be manufactured and supplied in one length on a suitable drum. No through joints will be allowed. All cables shall be of one manufacture.

Where individual cables are run on the surface suitable supports shall be fitted to give a minimum clearance of 15 mm between cables and face of structure. Where cables are installed vertically, the cable shall be gripped firmly by clamps of an approved pattern.

Where cables are grouped and run on the surface they shall be carried on wrought-iron brackets or purpose-made clips of approved design, fixed at not more than 600 mm centres.

All PVC SWA cables run on the surface shall be adequately protected to a height of 2 m from the ground.

Where PVC-insulated SWA cables are laid in the ground they shall be laid on not less than 75 mm of sand, covered by a further 75 mm of sand and protected by means of continuous interlocking warning tiles of approved pattern.

All cable trenches shall be excavated to a depth of 0.5 m in unmade ground and 0.75 m where crossing roadways and backfilled by the builder who will also provide and install all necessary cable ducts and earthenware pipes for cable entry into buildings, but the electrical contractor shall be responsible for correctly marking out all cable routes, supplying and installing warning tiles and marker posts, and generally supervising all work in connection with the cable-laying requirements.

17. *Cable joints*
All cable runs between one definite terminal point and another throughout the whole of the installation shall be installed without intermediate joints.

18. *Light fittings*
All light fittings shown and listed on the drawings and schedules shall be provided

and installed. Fittings with non-standard suspension lengths shall be ordered to the correct lengths to suit mounting height as indicated on the drawings and schedules. The installation of light fittings shall include all necessary assembling, wiring and erection.

Terminations to non-pendant fittings shall be in heat-resisting flexible cord with porcelain-insulated terminal block connectors for connection to PVC-insulated cable or PVC-insulated PVC-sheathed cable.

Fluorescent fittings shall be mounted either directly or on suspensions from two BS conduit boxes installed at the spacing required to suit the fitting.

19. *Ceiling roses*
Ceiling roses shall be white of reputable manufacture in accordance with BS 67. They shall be of porcelain, or of plastic with porcelain interiors and shall be fitted with plastic backplates or plastic mounting blocks semi-recessed where necessary to comply with the I.E.E. Regulations.

Where they are of the three-plate type the live terminal shall be shrouded so as to prevent accidental contact when the cover is removed.

20. *Lamp holders*
Lamp holders shall be of the bayonet-cap type for tungsten lamps up to and including 150 watt, and of the Edison screw type for larger lamps.

Where they are integral with lighting fittings, they shall be brass with porcelain interiors. For use with flexible pendants, they shall be of white plastic with compression glands. Where batten lamp holders are installed, the lamp holders shall be of white plastic. In damp situations they shall be fitted with Home Office skirts.

Lamp holders for fluorescent tubes shall be of the heavy pattern bi-pin type of white plastic construction.

All lampholders shall be lubricated with molybdenum disulphide to ensure easy removal of threaded rings and lamps.

21. *Lamps*
Lamps shall be supplied and fitted to all points and fittings shown and listed on the drawings and schedules.

Tungsten filament lamps of 40 to 100 watt (inclusive) shall be of the coiled coil type. Lamps shall be pearl-finished when fitted in open shades or in globes which are unobscured and shall be of the clear type when fitted in closed units of opalescent glassware or any other type of fitting selected where the filament is not under direct vision.

The colour of all fluorescent lamps shall be the new white, 3500 K.

22. *Lighting circuits*
Lighting circuits shall be installed on the loop-in system with three terminal-type

ceiling roses with shrouded live terminal, integral backplate, earth terminal and break-joint ring. Looping shall not be carried out at switch positions.

No light fitting shall be connected directly to the hard wiring or have circuit wiring passing through it.

Cables on one circuit are not to run through the BS boxes behind ceiling roses or fittings on other circuits.

23. *Light switches*
Flush switches shall be rocker-operated with white flush plastic plates of the single-switch or grid-switch type.

Surface switches shall be heavy-gauge steel with conduit entries and shall have rocker-operated mechanisms. They shall have steel front plates of the single-switch or grid-switch type.

Switches outdoors or otherwise exposed to damp conditions shall be of industrial pattern watertight type with galvanized steel boxes and waterproof gaskets.

24. *Socket outlets*
All socket outlets shall be of the switched type with rocker-operated switch mechanisms.

Flush socket outlets shall be of the insulated pattern with white or ivory finish.

Surface socket outlets shall be metalclad type with steel front plate.

25. *Connections to space heaters*
The circuit to each unit heater, fan convector and other similar piece of heating equipment shall terminate in a double-pole isolator from which the final connection shall be made in heat-resisting rubber insulated cable in flexible conduit. The casing of the heater shall be bonded to the earth continuity conductor.

26. *Connections to motors and machinery*
The circuit to each machine shall terminate in an isolator as near the machine as possible. Where a motor starter is required it shall be placed adjacent to and immediately after the isolator. The final connection from the isolator or starter to the machine shall be in PVC-insulated cable in conduit. The rigid conduit shall terminate in a box approximately 300 mm from the machine terminals and the final section from this point to the machine shall be in flexible conduit. The metalwork of the machine shall be bonded to the earth continuity conductor.

27. *Regulations*
The installation shall comply with Electricity (Factories Act) Special Regulations, the Factories Acts, the Electricity Supply Regulations 1937 and any other applicable statutory regulations. It shall conform with the Institution of Electrical Engineers' Regulations for the Electrical Equipment of Buildings.

The installation and all material used shall comply with all relevant British Standards and Codes of Practice.

28. *Clearance from other services*
All electric conduit and equipment shall be installed at least 150 mm clear of any other metalwork, and in particular of any water, gas, steam or chemical pipes.

29. *Bonding and earthing*
All conduit connections, boards, fittings, trunking, etc., shall be properly screwed together so as to ensure perfect mechanical and electrical continuity throughout.

Great care is to be taken in bonding and earthing the installation and tests are to be carried out as the work progresses to check the electrical continuity of all metalwork, conduits, etc., and earth continuity conductors. This is particularly important where work is built into the fabric of the building.

For the purpose of estimating the electrical contractor may assume that he can earth to the supply authority's earthing terminal.

The electrical contractor shall contact the supply authority at an early stage in the works to ensure that a suitable earthing terminal will be provided.

The electrical contractor shall be responsible for the bonding and earthing of all exposed metalwork, structural or otherwise, and of the metalwork of any gas or water service, to the earthing termination at the intake position, in accordance with I.E.E. Regulations.

No earth continuity conductors shall be less than 4 mm^2 copper cable insulated and coloured green.

The steel wire armouring of the sub-main cables shall be efficiently bonded together and to the respective switchboard, distribution board, sealing chamber and conduits at which they terminate and to all adjacent metalwork.

The frames of all electric motors and starting panels, etc., are to be efficiently earthed. Where flexible metallic conduit is used, a stranded insulated and coloured green copper cable of not less than 6 mm^2 is to be run from the terminal box through the flexible metallic conduit to terminate in the first cast metal box in the conduit run.

The earth wire shall be attached at each end by means of a brass sweating socket, brass screw and spring washer.

Pipelines, tanks, vessels and all other equipment associated with the piping or storage of highly inflammable materials shall be statically bonded to an effective earth conductor by means of 30 mm x 10 mm hard drawn tinned copper tape, secured by means of the flange bolts. The earth conductor shall be taken to an earth electrode and bonded to it.

30. *Earth electrode*
Where an earth electrode is required it shall take the form of extensible copper earth rods driven into the ground at suitable spacing. The number of rods and

the depth to which they are driven shall be determined according to the soil resistivity at the site to give an earth resistance not exceeding 0.5 ohms.

Concrete inspection covers shall be provided over every earth electrode and a means shall be provided for disconnecting the bonding cable from every earth electrode.

Connection to and between electrodes shall be carried out in insulated stranded cable.

31. *Testing*

Continuity and insulation tests shall be carried out during installation.

At completion polarity, bonding, earth loop impedance, continuity and insulation tests shall be carried out on the entire installation and in each part of it. The tests shall be witnessed by the consulting engineer and shall be carried out in accordance with the requirements of the Institution of Electrical Engineers' Regulations. A completion certificate as prescribed by the I.E.E. Regulations shall be provided.

32. *Circuit lists and labels*

At each distribution board a circuit list shall be supplied and fixed on the inside of the distribution board door.

The list shall state clearly the position, number and wattage of lamps, socket outlets, etc., which the fuseways control. A sample circuit list shall be submitted for approval before installation.

On the cover of each distribution board, fuse switch, switch fuse, isolating switch and starter a 45 mm x 20 mm traffolyte label (white-black-white) shall be fixed and engraved in 5 mm characters, giving details of the service position and phase etc. In addition, traffolyte (white-red-white) labels engraved in 8 mm characters '415 volt' shall be fixed to all TP & N distribution boards.

All labels shall be fixed by means of four 4 BA round-headed brass screws.

33. *Identification of cables*

All power, instrument, control and indication cables shall be provided with indestructable cable marking collars which shall bear the cable number. The marking collars shall be fitted at every cable termination.

The individual cores of cables shall be numbered to indicate which terminal they are connected to.

34. *Fuses*

HRC fuse links of the current rating shall be supplied and installed in all fuse carriers.

Bibliography

Amongst the many books published on this and related topics, the following are suggested for readers at this level

Donnelly, E. L. (1972)   *Electrical Installation, Theory and Practice*
                                                Harrap
Francis, T. G. (1971)   *Electrical Installation Work*
                                                Longman
Henderson, S. T. and   *Lamps and Lighting*
  Marsden, A. M. (1972)   Arnold
Ibbetson, W. S. (1970)   *Electric Wiring, Theory and Practice*
                                                E. & F. N. Spon
Jay, P. and Helmsley, J. (1968)   *Electrical Services in Buildings*
                                                Elsevier
Johnson, R. C. (1971)   *Electrical Wiring: Design and Construction*
                                                Prentice-Hall
Miller, H. A. (1965)   *Electrical Installation Practice*
                                                Arnold Technitrade
Neidle, M. (1970)   *Electrical Installation Technology*
                                                Butterworth
Pritchard, D. C. (1978)   *Lighting*
                                                Longman
Raphael, F. and Neidle, M. (1974)   *Electric Wiring of Buildings*
                                                Pitman
Steward, W. E. and Watkins, J. (1976)   *Modern Wiring Practice*, 8th edition,
                                                Butterworth
Strakosch, G. (1967)   *Elevators and Escalators*
                                                Wiley

Index